POPULAR MEDICINE

I0487459

RADIATION AND HEALTH

(SECOND EDITION)

NELI MELMAN

MEMORIES OF VICTIMS OF RADIATION

In December 2003 the General Assembly of the UN has supported the decision of Council of heads of states of the CIS on a declaration on April 26 in the International afternoon of the victims of radiation accidents and accidents

Splitting of the atom has changed the whole world how we think of him. And it slowly goes on the way, to not having equal, accident.

Albert Einstein

Radiation accident is a radiation accident somewhere is a catastrophe everywhere.

(Aphorism)

Dear readers!
If to you it is unfamiliar what or a word, please, address "The dictionary of terms and concepts" at the end of the book.

TABLE OF CONTENTS

INTRODUCTION

The end 19 and the beginning left centuries, as well as the next decades of the 20th century, were marked by the greatest opening. With good reason treat them opening the X-rays and the splitting of an atom.

So - called X-rays were a casual find of the genius German scientist William Conrad Roentgen (1845 - 1923).The splitting of an atom has defined the science birth – nuclear physics.

As a result of the splitting of the atom, the new type of radiations called the ionizing radiation is received. It is common it indicates the term "radiation". The specified opening has predetermined unpredictable changes practically in all areas of life of mankind, including human health.

The complexity and diversity of problems of radiation have caused need of allocation of new scientific and applied specialties - radio medicine, radiation oncology, radiobiology, radio physics, radiochemistry, radio ecology, radio power, radio veterinary science and many others. Epoch-making opening and the subsequent) development of a multidimensional problem of radiation is noted by high National and International awards, degrees, ranks.

It is difficult to overestimate names of the largest scientists in permission of many problems of radiology. For this scientist about 30 erudite physicists, chemists and biologists have conferred the Nobel Prize.

Among them, there is Antaean Becquerel; spouses Pierre Curie and Maria Sklodowska - Curie opened the spontaneous radioactivity (1903).

Receiving new radioactive elements was the basis for the award of the Nobel Prize to spouses Frederic Joliot and Irene Joliot-Curie (1935). Enrico Fermi has conferred the Nobel Prize for a new way of receiving radioactive elements (1938). Ernest Lawrence became Nobel Prize laureate for the invention of a cyclotron (1939).

The atomic reactor for the first time in the world was designed in the USA by the Nobel Prize laureate Enrico Fermi (1942). The Nobel Prize laureate in the field of medicine became the first woman Rosalyn Yellow, applied radioactive substances to diagnostics of a number of diseases (1977).

It is difficult to overestimate the value of scientist in permission of many problems of radiology and names of other Nobel laureates.

It is necessary to remember that a number of outstanding scientists worked in very difficult conditions, were pursued because of nationality and political convictions. They sacrificed a lot of things for the sake of The decisive turn in studying of various aspects of the influence of the ionizing radiation on the person in many respects is caused by two tragic events of the left century which have received the international resonance. The mankind for the first time has seriously thought of a question of the survival.

Treat these events:

- atomic bombing of the Japanese cities of Hiroshima and Nagasaki (August 1945);
- accident at the Chernobyl nuclear power plant (April 1986, Ukraine, the former Soviet Union).

It is necessary to emphasize that the accumulation of atomic potential in the military purposes which has begun in 40-50 years of the 20th century and continuing so far is very dangerous to mankind. To a huge regret, it is impossible to be sure that use of atomic weapons will never repeat.

The accruing threat of radiation terrorism which has begun in 80-90 years of the left century, in turn, for a long time has defined the relevance of studying of various aspects of a multidimensional problem of radiation.

And, at last, wide use of atomic energy in numerous areas dictates need of comprehensive study of this problem, in particular, of influence on the health of the person.

What main directions of scientists of the influence of the ionizing radiation on the health of the person?

It should be noted that results of the conducted scientist aren't always unambiguous. It, in a certain measure, is connected with the continuous change of methodology and a technique of, scientists, their opportunities, and many others. Besides, consequences, in particular medical, are defined by many characteristics of radiation pollution, and influence.

Throughout the almost century the main directions of the scientist of the influence of radiation on the person include:

- studying of mechanisms of radioactive effects;
- definition of weight of damages depending on many characteristics of radiation (character of a source, a dose, radiation duration, etc.);
- studying of the damaging action of radiation at various level (body cellular, sub cellular);
- improvement of diagnostics and treatment of radiation damages;
- primary and secondary prevention;
- methods of medical and social rehabilitation;
- methodology of long-term observation over victims and their descendants.

The purposes and problems of the listed scientists are concretized depending on the reasons and many other characteristics of radioactive effects.

Very great value is attached to the establishment of interrelation and integration of various numerous National and International institutions into the organizations of special services and training of specialists of all levels. For these scientists and their introduction in practice are allocated, in particular to the USA, huge allocations of both the state and private funds. Already it is possible to say that in many directions of researches certain positive results are received. It is much made but remains the questions which are subject to urgent, deep and comprehensive study even more.

Studying and other problems at higher methodical level with the participation of experts of many profiles are in addition begun with the beginning of 21 centuries radio physics, radiobiologists, mathematicians, doctors and others).

Treat these problems:

- biodosimetry and creation of new automatic systems of observation, diagnostics, treatment, and prevention which were injured from radiation;
- receiving new radio protectors and other methods of prevention, and treatment (antibiotics, etc.) at ray influences and diseases.

The Chernobyl accident has expanded and has deepened new chapter of radio medicine – studying of long impact on the person of small doses of radiation. This very complex and current problem borrows, and will long occupy one of the central places in versatile scientist about the damaging action of the ionizing radiation per the person.

A comprehensive study of medical consequences of the accident on the NPP Fukushima in Japan will be the new section.

In an experiment studying of mechanisms of radioactive effects, their diagnostics, treatment, and prevention continue.

In the fiftieth years of the 20th century the prominent scientist disgraced academician Audrey Dmitrievich Sakharov in the Soviet Union has been concerned by the fact that small doses of the ionizing radiation can have the damaging effect on a human body. Similar fears were expressed and other prominent scientists state.

Long-term experience and complex scientist demonstrate that hopes of mankind for "peaceful atom" far was short equaled. It means that achievements in many fields of science and technology can become fraught with the serious accidents and accidents exerting the huge negative impact on the health of the person. Therefore one of the main tasks of State - participating in atomic projects, occupies safety issues of sources of artificial radiation.

The scientists devoted to the damaging influence of natural sources of radiation are intensified (space beams and others). Results of the conducted scientist are regularly discussed at numerous local, National and International meetings. The received results are analyzed and generalized, decisions on further researches and practical recommendations are made. It should be noted that a number of scientific recommendations, in particular in Ukraine, Belarus, and Russia, don't find support in Governmental bodies. It generates various forms of a protest (demonstrations, strikes, hunger strikes, etc.).

Already today the number of scientific publications on problems of an influence of radiation on the person is millions. It is counted that only mentions of the Chernobyl tragedy approach three million.

Questions of radioactive effects on the person concern world community.

The word radiation has deeply entered the consciousness of people as an image something terrible. It finds reflection in fiction, art, theatrical performances and movies. Unfortunately, the considerably smaller significance is attached to increase in medical literacy of the population. Editions for the population on a problem of influence of radiation on health are presented, mainly, in separate publications, newspaper articles. Besides they are sometimes written not by experts. This information isn't always available to a wide range of the readers including who have undergone beam influence from different sources (radio treatment, accidents, radioactive waste, etc.).

Questions of radiation medicine are practically absent in plans of training of doctors and other health workers.

We sought to meet this lack and to give to a wide range of readers an idea as of radiation problem in general, and consequences of her impact on the health of the person, methods of prevention and treatment, an active position of the population.

The basis for a reprinting of the real grant was important results of the continuing scientists.

The book is written on materials of the analysis of authoritative sources of the world literature and own experience of work in the field of radiation medicine.

GENERAL IDEAS ABOUT RADIATION

Everything in the world consists of the smallest particles called by molecules and atoms.

The molecule is the functioning unit consisting of two or more atoms.

Atom represents a particle of elements which can enter chemical reactions. His name has come from the Greek word "atom" "-indivisible. Such representation of Ancient Greek philosophers reflected in his name is disproved by modern science. It is proved that atom is divided into smaller particles. Nevertheless, the term "atom" was fixed and remains.

Molecules and an atom aren't visible with the naked eye. The size of the atom is equal to one billion centimeters. On the end put to the end of the offer, 100 billion atoms can go in.

The atom structure in a miniature reminds solar system.

Center of the atom, there is a positively charged kernel which is the carrier of its properties. The kernel in 100 thousand times is less than the sizes of the atom. It consists of protons and neutrons.

Atom splitting (atomic disintegration)

The atomic nucleus of cells of tissues of the person contains chromosomes. These are double spirals of DNA with a set of genes carriers of a genetic (hereditary) code.

Around a kernel on the closed orbits particles which are called electrons rotate. The number of electrons in the atom is equal to a number of protons in his kernel.

The vast majority of atoms of various chemical elements of the environment are stable, i.e. not be exposed to causeless changes as a result of which there is an ionizing radiation (from the Latin word radius, "radiate" - to radiate.

Along with it, in nature, there are substances which atoms are unstable (for example, uranium). They spontaneously collapse and turn into other elements, allocating at the same time radioactive rays (alpha, beta, gamma).

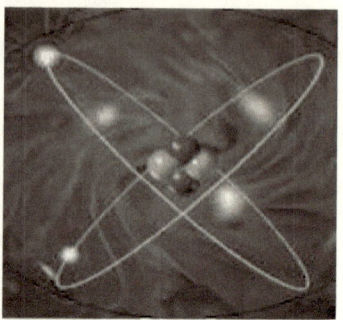

Scheme of a structure of atom

Atom splitting (atomic disintegration)
The atomic nucleus of cells of tissues of the person contains chromosomes. These are double spirals of DNA with a set of genes carriers of a genetic (hereditary) code.

Around a kernel on the closed orbits particles which are called electrons rotate. The number of electrons in the atom is equal to a number of protons in his kernel.

The vast majority of atoms of various chemical elements of the environment are stable, i.e. not be exposed to causeless changes as a result of which there is an ionizing radiation (from the Latin word radius, "radiate" - to radiate.

Along with it, in nature, there are substances which atoms are unstable (for example, uranium). They spontaneously collapse and turn into other elements, allocating at the same time radioactive rays (alpha, beta, gamma).

Radionuclide:	Half-life period	Type of radiation
Americium-241	432.2 year's	alpha - weak gamma
Caesium-137	of 30 years	beta, gamma
Caesium-134 2	year's	gamma, beta
Iod-129	1.5 million years	beta
Iod-131	8 day's	beta
Iridium - 74	days	beta, gamma
Plutonium-238	80years	alpha
Plutonium-239	24.000 years	gamma
Strontium-90	29 year's	gamma
Strontium-89	50 days	gamma

Table 1. Half-life period and character of radiation of some radio nuclides.

It is important to emphasize big fluctuations of a half-life period for various substances - of several days (iodine-131) to hundreds and thousands of years ((americium-241, plutonium-238, iodine-129, and others).

Isotopes of the same chemical element can differ in quantity of neutrons and duration of a half-life period (for example, iodine, and strontium). Some isotopes let out various beams. So, for example, caesium-137 is let out by a beta - and gamma beams, americium-241 – an alpha - and weak gamma rays.

Radiation types (α, β, γ) are characterized by the release of a certain amount of the energy having unequal penetration. Therefore their impact on a live organism is very various.

Alpha (α rays) are formed at the disintegration of atomic kernels of uranium, thorium, radium, plutonium and other chemical elements. They bear a large amount of energy. Their ionizing ability not really big. These rays are late the sheet of paper and practically can't get to an organism through the intact skin and mucous covers. Hit in an organism is possible through an open wound, with the inhaled air and food. In such situations, the specified beams become very hazardous to health and human lives.

Beta (β rays) result from radioactive decay of a number of elements, for example, cesium. They move with very high speed and are dangerous both for the hit on the surface of the skin and in inside organism. The penetration to beta beams is higher than alpha rays. Take place in body tissue to the depth of 1-2 cm.

Gamma (γ rays) have the shortest wavelength and have the biggest penetration spread with the velocity of light. The thick lead or concrete board can protect their penetration into an organism.

The name alpha, a beta - and scale - beams are offered the Nobel Prize laureate Ernest Rosenfeld (1871 – 1937). X- rays called by the name of their pioneer on the properties they are close to gamma rays.

The penetration of the listed rays is inversely proportional to the length of their rays. In other words, then rays length is shorter, that the penetration is more.

Main characteristics of radiation

Radioactive radiation has a number of certain characteristics. They can conditionally be divided on quantitative (X-ray, Sievert, etc.) and the high-quality, reflecting character impacts, in particular, into the person. Radiation types and their penetration rays length are shorter, that the penetration is more.

Quantitative characteristics of radiation

The dose or its activity and power characterize the quantity of radiation. They are defined and say in below the presented international units.

Cur- international traditional unit of measure of radioactivity equal to the radioactivity of 1 g of pure radium. She reflects the rate of decay of radioactive material. It is called in honor of world famous erudite Maria Sklodowska Curie.

Becquerel (Bq) - the standard international unit of radioactivity equal to one disintegration in a second. In the fact, it too that Curie only in system" C". The unit is called in honor of the famous physicist Nobel Prize laureate Antoine Henri Becquerel.

Rad- unit of measure absorbed, i.e. absorbed, radiation doses.

Gray (Gr) unit of the absorbed dose in the C system, equal 100 Rad.

Rem - (Rem - an X-ray equivalent) the unit of the measure giving an idea of the damaging action of radiation. Usually, she is transferred to Sievert.

Sievert - (Sv) a unit of measure of the equivalent absorbed dose, equal 100 Remes.

Ber- outside the system unit of measure of an equivalent dose of radiation (biological equivalent of X-ray).

X-ray (Roentgen) - unit of measure of the absorbed dose.

Use of that or other unit is defined by a research object (the person, the environment, etc.)

Star- used infrequently in various

Names and the size of quantities of radiation change, new are periodically offered. Now for determination of a size of radiation units Sievert, and Gray are most often used.

Quantitative indices are widely applied to the establishment of extent of radiation contamination of foodstuff, water, the soil, construction materials, raw materials, industrial output, etc. Use of the same units of measure of a dose of the ionizing radiation improves mutual understanding of researchers and workers of practice.

Besides the listed indicators, the influence of the ionizing radiation is estimated also by many other parameters (the name and number of isotope distance from a source, radiation duration, etc.).

Methods of definition of the amount of radiation

Radiation has neither color nor a smell. Therefore her definition is possible only by means of the special devices called radiometers (dosimeters). It is impossible to overestimate the value of determination of radiation level in various environments (the soil, air, water, etc.) and live organisms, in particular, at the person.

The field of physics which is engaged in the determination of quantity, intensity, and distribution of the ionizing radiation in space and time is called dissymmetry.

The dose characterizes the size of the absorbed energy. Determination of radiation level in a human body includes both the general radioactivity and her level in separate tissue and bodies (thyroid gland, bones, etc.).

All methods of dissymmetry can be divided into three main groups: tool, and settlement. Devices for the definition of radiation are called radiometers, or dosimeters. Their design and the sizes are very different.

Direct measurement of the size of radiation absorbed by living tissue is impossible. In this regard, direct measurement of a dose is carried out in certain physical or chemical environments. They are called dissymmetric model environments – detectors. Liquid or gaseous and firm environments are for this purpose used. Designs of the various detectors differ in sensitivity degree to radioactive materials. Use like the detector is also defined by the structure of radioactive beams (x-ray and scale – rays, or an alpha - and beta rays) and their intensity.

Dosimeters are issued for individual use, the scientific purposes, military conditions, etc. They are stationary and figurative. An object of a research determines the size, a form and a design of the used dosimeter.

For measurement of the power of a dose the beta - and gamma radiations are used three types of dosimeters – low background average and high ranges.

Work with dosimeters is in a field of activity of experts - meter men.

Improvement of dosimeters belongs to current problems of radiobiology and radio medicine. Representatives of different specialties work on the permission of the specified problem (physicists, chemists, etc.).

Not always it is obviously possible to determine in due time sizes of individual and collective doses of radiation. Their definition is of great importance for holding rational organizational and treatment-and-prophylactic actions and in the subsequent – diagnostics and the forecast of damages.

Various types of dosimeters

In this regard, throughout the last four - five decades, methods of retrospective assessment of the saved-up doses are developed. So, the American and Japanese scientists, on the basis of special physical and mathematical calculations, have developed the DS86 system. By means of this system radiation doses at the persons who have survived after the atomic bombing of the Japanese cities of Hiroshima and Nagasaki are calculated.

Recently are offered and new methods of physical and biological dosimeter are already applied to the retrospective assessment of the saved-up doses. They are called reconstructive (restored). One of perspective techniques is the spectroscopy of an electronic paramagnetic resonance of enamel of teeth (EPR dissymmetry). The specified technique is used at inspection of victims after the Chernobyl accident. According to experts, she is the most sensitive and specific biological indication of a dose of radiation among all methods. Very important and the fact that its application isn't limited to past tense after radiation.

In the conditions of specialized laboratories definition of the absorbed radiation of plutonium and strontium according to their content in urine or in fecal, and at the feeding women – in breast milk is possible.

Scientists of Ukraine have offered a technique of retrospective assessment of radioactive effects on various objects (the soil, plants, etc.). The test glycophorin is developed to determine the damaging influence of radiation on the human genome in Livermore National Laboratory of the USA. He allows reconstructing the received and saved up doses of radiation and is predictive for determination the probability developing of cancer.

The cytogenetic method (FISH) allowing to reconstruct a dose at the acute and chronic influence of radiation and also to predict the adverse effects induced by radiation is used.

The possibility of a definition of radioisotopes in an organism on the basis of studying of their markers is studied.

The international Program for medical consequences of the Chernobyl accident has offered the model for reconstruction of doses of radiation of thyroid gland radioactive iodine in those places in which direct measurements weren't taken. Simple methods of the blood test are developed for the definition of a damage rate of tissues of the person in connection with radiation therapy, the American scientists. They need further studying.

The research of a condition of a genome according to a definition of chromosomal aberrations, it is perspective, changes in their structures.

Systematically one method is succeeded by others. Scientist in this area intensively quantities.

Qualitative characteristics of radiation

The qualitative characteristics of radiation used in connection with concrete situations of radioactive infection are listed below.

Quantities of ionizing radiation -the amount of the energy absorbed in the unit of mass of environment.

Absorbed quantities - the amount of the energy absorbed by the unit of mass of the irradiated body.

Equivalent quantities -the biological activity of different types of radiation (alpha, beta, scale etc.), is expressed in Sievert or Ber. She equals to the work of the absorbed dose of this type of radiation multiplied by the corresponding coefficient of quality. The coefficient is equal 1for the x-ray - scale - beta ray; 10 – for alpha rays; 20 – for neutron radiation.

Power of quantities allows predicting results of radiation exposure of the person. In each case, it equals to the dose received by a unit of time (second, hour).

The effective (equivalent) quantities reflect the total value of the dose received by the radiation source by different bodies and fabrics according to their sensitivity to the action of radiation. In places of radioactive pollution, for example, after the accident et the Chernobyl nuclear power plant, the annual effective dose which is expressed in µSv in a year is defined.

Specific activity – the indicator which is the criterion of impurity of foodstuff, waters, soils, construction materials, raw materials and industrial products.

Therefore, qualitative characteristics of the ionizing radiation allow gaining an impression about the nature of her impact on the person and the environment. The list of qualitative characteristics of radiation all the time is replenished.

Quantitative and qualitative characteristics are complemented.

SOURCES OF RADIATION

Distinguish two groups of sources of the ionizing radiation – natural and artificial. The numbers of sources of natural radiation are much less than artificial.

Sources of natural radiation

Radiation has appeared earlier, than the earth and life on it. The radioactive splitting happening in a solar sphere and stars was and remains her source. Radiation the most ancient factor of the environment which has arisen 20 billion years ago.

Life on the earth develops against the background of natural radiation. All history of mankind passes in an environment. Radioactivity same factor of the environment as beams of the sun and stars, fluctuations of temperature and atmospheric pressure, change of time. It is her latest, quality and "natural", "background" radiation is designated by terms.

In Russian-speaking literature the term "normal radiation." is more often used in the light of an environment, latest, recent, of scientific data the specified term is unauthorized since natural radiation under certain conditions can have the damaging effect on an organism of the person.

Three sources of natural radiation are known.

- Space rays.
- The radioactive elements which are formed in the atmosphere.
- The radioactive materials which are in the earth ("terrestrial radiation").

Space rays contain protons, electrons, x and gamma rays. They make 21% of natural radiation.

The considerable part of sunrays is absorbed by the atmosphere, creating one more constant source of radiation.

Radioactivity of sources of the earth (radon, thorium, potassium, etc.) much more (79 %.). The main source of radioactivity of the earth is the radon which is in earth subsoil (55%). Much less than a percent falls to the share of space beams and radioactivity of the land surface.

Trace quantities of radioactive elements (potassium-40, lead-210, etc.) contain in a human body. They get to an organism with food and the inhaled air.

Sources of space beams as it is stated above, this sun and stars. Their level isn't identical in various geographical zones of the globe.

According to the data of the American National Council of radiation prevention, during 1945-1995 radiation of space beams above sea level made 40 μ/X- rays in a year, above the ground – 55 μ/X- rays in a year.

Space rays can constitute a certain danger to pilots and, to a lesser extent, for passengers of planes. In an hour of stay in the air of people receives 0.5 μ/ Rem. Much attention in the USA, Canada, and the European Union countries is paid to this question. Admissible doses of radiation for crew and passengers of planes are established.

Impact of space radiation on astronauts is studied. It is noted that risk of her influence on female astronauts in 2.5 times more than on men.

The radiation background of air, water, and the food is characterized by big fluctuations, making 20-400 μ/X- rays.

The radiation background in states of our country is non-uniform.

The average level of radiation for the resident of the USA generally is defined by natural sources (82%). Only the small percent of annual radiation is caused by sources of artificial radiation of-18%.

Since 1992 the annual dose of radiation exposure has increased and makes 360 Rem/year. It is calculated what in 70 years of the life of people receives 17 X-ray.

Therefore, in usual conditions of the USA sources of natural radiation of the person are four times more, than artificial.

Unusually high radiation background (0.5-1.2 Rem/year) is registered in some places of Sri Lanka, India, Brazil, Iran, and Italy. May be, it is connected with big bedding of uranium ores and an exit to a surface of redoing sources. Such background is the risk factor of developing lung cancer.

Uranium and its derivatives (thorium, radon, etc.) as a source of natural radiation constitute a certain danger for working in mines, especially at non-compliance with preventive measures and safety measures. In conditions of annual radiation equal 0.3-0.6 μ/Sv 3% - 1 μ/Sv and more than 1.5%-is higher than 1.4 Stars.

According to the Russian scientist of 90% of the population lives in conditions of the raised radiation.

During the life of hundreds and thousands of generations of living beings, practically all have adapted to activity in the conditions of a natural radioactive background and its local fluctuations. As it is stated above, during all life the human body contains trace quantities of some radioactive elements. Centuries-old evolution of mankind has defined development in an organism of protective mechanisms, including from radioactive effects.

For the last decades, the level of natural radioactivity in many places of the planet has increased. It is connected with the wide use of radioactive materials in many industries, power engineering specialists, medicine, and also for the military purposes.

The maximum permissible dose for the persons working with radioactive materials makes 5.000 μ/X- rays in a year. In the USA the radiation background in a radius of 50 miles from nuclear power plants shouldn't exceed 0.01 μ/Rem. Similar standards, are regulated practically for each country.

The radiation background is systematically controlled near nuclear power plants, cyclotrons, and places of the test of atomic weapons, submarines and burial of radioactive waste.

Data on the negative influence of small doses of radiation on health were the basis for studying of a possibility of a decrease in the level of natural radiation.

Sources of artificial radiation
It is the incomparable bigger negative impact on the person can make sources of artificial radiation, i.e. received by splitting of atoms. It is connected not only to their large number. It is established that artificial radioisotopes I differ in more long half-life period, a tendency to spontaneous disintegration and bigger penetration

After the invention in 1935 of a cyclotron, the American erudite D. Hedwig (1891-1974) has continued works on splitting of an atomic nucleus for the scientific purposes (in Moscow area Obninsk, the former Soviet Union; in the State of California at Berkeley's university - E. Laurens with employees and others). The quantity of methods and ways of receiving artificial radiation in many countries steadily increases. The quality of installations for her receiving, radiation protection of personnel and the environment is improved.

It is advisable to differentiate sources of artificial radiation, but not to unite them the general heading "Technological".

Artificial radiation sources are extremely various according to the many characteristics and impact on the person. There is no need to prove distinction of sources of radiation used for example for diagnostics, with accidents on nuclear power plants, etc. Now it is possible to allocate the following sources of artificial radiation which can have the damaging effect on the person. Treat them: use of radiation in medicine, nuclear power plants, production, the military purposes, radiation waste.

RADIATION AND MEDICINE

General information
The opening of X-rays and splitting of an atom were a push for studying of their influence on the health of the people.

Pierre-Curie and Maria Sklodowska - Curie's pioneer opening have made possible practical application of radioisotopes in medicine.

In 1901 in France the isotope discovered in the country for treatment of the cancer patient is for the first time used.

Three years later the dermatologist from Melbourne (Australia) has applied isotope Polonium at the patient with cancer of the skin.

The name of the isotope is offered in honor of M. Sklodowska - Curie's homeland.

Only after 1924 intravenous administration of products of radioactive radium with the diagnostic purpose is begun. The radiation medicine began to develop especially successfully since the end of the 50th years of the left century. This beginning is connected with works of the first woman of the Nobel Prize laureate in the field of medicine of the American Rosalyn Yalou and the employees. They for the first time in the world have applied radioactive iodine -131 to the diagnosis of diseases of the thyroid gland. There are messages about the death of 336 people participating at the beginning of the left century in various powered by scientist with radioactive materials. Cases for a disease of cancer at the doctors participating in treatment of victims of the accident on the CNPP and the persons lighting everything there the events are known (cameramen, journalists, etc.).

From many casual observations of centenary prescription concerning the development of tumors in the staff of similar institutions before long-term scientific research – such is a way of a new division of science and practice of radio medicine. At the beginning, the specified division of science included only influence studying. In the next years, a subject of radio medicine was defined opportunities of use of radiation for diagnostics and treatment. The doctors working in the field of radio diagnosis and radio treatment systematically and closely cooperate with radio physicists, radio chemists, radio biologists, ecologists and other experts.

Since the beginning of the left century and so far radiation methods of diagnostics and treatment increase progressively and improved. From 1980 to 2007 their use has increased more than in 60 times. Today it is already impossible to imagine and it is impossible to underestimate the value of these methods in diagnostics and treatment of a widespread and serious illness. Radiation methods are applied in cardiology, pulmonology, gastroenterology, neurology, obstetrics and gynecology, urology, nephrology, orthopedics, etc. In fact, there is no medical area in which radiation methods wouldn't be applied.

For the medical purposes gamma rays of short-lived isotopes are used now (iodine, bismuth, cobalt, and others). They can be entered intravenously, by inhalation, and through digestive tract. Isotopes for the medical purposes receive in special cyclotrons.

Radiation methods allow diagnosing early stages of cancer and its metastases, to define a blood-groove in coronary vessels of heart, kidneys and other bodies. The value of such methods in the assessment of the function of heart, lungs, a liver, kidneys, thyroid gland, etc. is high. So far in the developed countries use of radiation diagnostics makes 1.9% of all diagnostic methods.

It should be noted expressed fluctuations of beam loading when using various methods of radiation diagnostics.

For example, at an X-ray analysis of extremities, it makes 0.01µ/Sv, and scanning of thyroid gland – 14 µ/Sv.

It should be noted that similar indicators in many respects are defined by the quality of the used equipment. The equipment, the fever rays loading are more perfect. Comparison radiation diagnostics are of interest load with a natural radiation background.

So, at a computer tomography of a stomach, it is equivalent to three-annual loading, and the heads – eight-months.

Apply the isotopes with a short half-life period i.e. which are quickly collapsing to the diagnostic purposes. At the same time, their primary accumulation in that or other body is considered. So, for example, for diagnosis of diseases of the thyroid gland, it is used iodine1.

The number of diagnostic techniques steadily increases, decreases loading at their use, prevention and treatment of possible complications are improved.

One of the most perspective techniques is the pharmaceutical diagnostics allowing studying a condition of cages of that or another body. Other high-informative methods with a rather small beam loading also are in process of approbation. Some of them can be applied during pregnancy. Introduction of some the x-ray contrast of substances isn't recommended to the women nursing (iodine and others).

Everything stated, undoubtedly, confirms big achievements of modern science and medical practice. Unfortunately, their wide use is inaccessible to many economically undeveloped countries.

Radiation treatment (radiotherapy.)

Incomparably more beam loading at the treatment of various diseases, than for their diagnostics.

Annually in the world radiation methods of treatment receive 18 million patients. It makes 10% of a number of all other methods of treatment of much serious and widespread illness.

About 10 methods of radiation treatment are known. Their number increases. Depending on an arrangement and microscopic structure of cancer, radioactive materials are entered in various ways (external, intravenous and others).

Radiation treatment is used at cancer of various localization and structure. It is shown to patients with cancer of a brain, chest gland, uterus, throat, lungs, a pancreas, a prostate gland, skin, a backbone, a stomach, soft fabrics, by leukemia, a lymphoma. At some forms of cancer, after successful operation or chemotherapy, radio treatment is carried out with the preventive purpose, i.e. for prevention of repeated emergence of a tumor and metastasis. Besides, radiation therapy can be appointed for reduction of the pain caused by metastasis of cancer and other reasons.

The persons who are carrying out radiotherapy (doctors, nurses, laboratory assistants), are under regular observation which purpose is the timely effect of radiation influence on their health.

In the USA the Committee on the regulation of doses of radiation strictly regulates her admissible doses. For the medical purposes, they have to make no more that $0.1~\mu/Sv$ in a year, for working with radioactive materials – $50~\mu$ / Stars in a year.

Carrying out difficult surgeries under control of radiological diagnostic units are represented promising (for example, brain operations).

Enough the method of the directed irradiative effects directly on a cancer tumor was widely adopted that considerably reduces beam load of the patient.

In respect of decrease in beam loading, radio pharmacological treatment is perspective. This method allows delivering by means of isotopes drugs directly in cancer cells of that or another body.

A side effect of radiation treatment nevertheless remains. Many symptoms of radioactive effects (change in the system of blood formation, digestive tract and nervous system) are reversible. However, some complications can constitute the health hazard. This problem constantly is in sight of doctors (immunologists, hematologists, etc.).

The persons who are carrying out radiotherapy (doctors, nurses, laboratory assistants), are under regular observation which purpose is the timely effect of radiation influence on their health.

In the USA the Committee on the regulation of the doses of radiation strictly regulates her admissible doses. For the medical purposes, they have to make no more than $0.1~\mu Sv$ in a year, for working with radioactive materials – $50~\mu$ / Stars in a year.

Carrying out difficult surgeries under control of radiological diagnostic units are represented promising (for example, brain operations).

The method of the directed radioactive effects directly on a cancer tumor was widely adopted that considerably reduces beam load of the patient.

In respect of decrease in beam loading, radio pharmacological treatment is perspective. This method allows delivering by means of isotopes drugs directly in cancer cells of that or another body.

A side effect of radiation treatment nevertheless remains. Many symptoms of radioactive effects (change in the system of blood formation, digestive tract and nervous system) are reversible.

However, some complications can constitute the health hazard. This problem constantly is in sight of doctors (immunologists, hematologists, etc.).

The high efficiency of radiation treatments, the relatively small complexity of the application determines the need of improvement, primarily to reduce side effects.

INDUSTRIAL SOURCES OF RADIATION

Connected radiation with the production conditions, it is worth defining the term "industrial (professional) exposure".

In what industries use radioactive substances? They are automotive, aircraft, construction, particularly residential homes, fertilizer, fossil fuels, electronic appliances, road construction, and testing of oil and gas, cable construction, printing and many other industries. The list is regularly updated and will be updated in the future. The degree of radiation of different materials was designated so-called "characteristic of the radiation characteristic". Until recently, it was used three symbols, reflecting the nature and site of radiation.

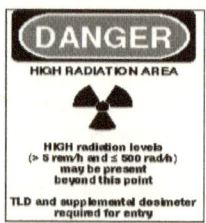

Danger area

Clarification zone

ATTENTION! Radiation zone

In 2005 the World organizations have offered a new symbol of radiation pollution.

New symbol of radiation pollution

Despite criticism, a symbol it is accepted.

The international organizations of control of radioactive infection mark out three categories of technological sources. The division into categories is based on the level of potential danger.

Category 1. The teletherapeutic devices irradiating installations.

Category 2. Stationary industrial sensors with highly active sources.

Category 3. Stationary industrial sensors with low-active sources.

Similar classifications are used for the organization of control of work and also drawing up protection programs and safety.

Due to the continuous expansion of radiation methods in the industry, including in the food industry, systematic control of their level is necessary. Not less important is a regular observation over personnel.

Use of radiation in many branches of the industry and scientific research promoted and promotes new achievements and progress. At the same time, it is worth to remember a possibility of her negative impact. "Peaceful atom" not always is "peace".

NUCLEAR POWER PLANT (NPP)

General information
Splitting of the atom has created a new source rather inexpensive and very powerful energy, including electric and thermal. The nuclear power began to extend quickly enough around the world. To a huge regret, a lot of things at the same time turn the heavy serious consequence for mankind and the environment. The emergence of nuclear power plants with evidence has shown the back of so-called peaceful atom. The fears about it at the beginning of the 20th century were expressed by *Julio Curie*:" Most of all concerns me who will use discovery which I had made".

A number of the International and National departments are engaged in nuclear power. Prominent scientists and authoritative institutions of various profiles are connected to their work. At once after the accident on Tokaymura and all subsequent time permanently and quickly in various countries discussion of an event is carried out, measures for accident elimination control of all existing NPPs are proposed. All events around the world are widely lit in mass media (radio, television, Internet). The situation in the Japanese nuclear power plants where the strong for the history of this country of an earthquake was resulted by several explosions has set other countries thinking about the future of the nuclear sector. Quite naturally that there are questions again: whether those risks which the mankind is forced to pay for use of atomic energy are justified? Whether can provide traditional sources sufficient electricity generation for the growing economies. A number of the states, for example, the USA, China, Russia, France, have stated that they won't refuse the nuclear power. Some states haven't come to the final decision yet and even periodically change it. The responsible answer to the matter requires the detailed analysis on the basis of the bigger number of data. However, some preliminary conclusions are drawn already now.

Generally they concern studying of a state and increase in a safety culture at all levels of management and regulation of nuclear power, the attentive relation to details, implementation of effective programs, identification, the analysis and elimination of the factors reducing safety, and ensuring effective management of knowledge in nuclear area, literacy of personnel. The need of improvement of the quality of training of shots for nuclear power is noted. For achievement of this purpose supplying countries of the NPP have to create the centers of training of specialists for the countries accepting nuclear technologies. A row new is supposed. Issues of non-proliferation of nuclear weapon and prevention of hit of nuclear materials and technologies in hands of terrorists are discussed. The Japanese nuclear crisis will be long and not knowing frontiers.

In the world 442 nuclear power plants which are an important power source in 32 countries are constructed. Their construction continues.

Especially the NPP in India, France, Belgium, Northern Korea, Switzerland, Sweden, Japan, Russia, and Ukraine is a lot of. They make from 36% to 73% of all sources of the electric power.

In the USA nuclear power plants give about 20% of percent of the electric power. At the same time it is necessary to emphasize, the high level of prevention of possible accidents. It is provided with the improvement of a design of nuclear reactors. Also, methods of protection of employees and the environment are improved.

In the normal mode of the NPP are safe, but emergencies with emissions of radiation exert the destructive influence on ecology and health of the population. Despite the introduction of technologies and the automatic systems of monitoring, the threat of emergence of the potentially dangerous situation remains. At each tragedy in the history of nuclear power own unique anatomy. The human factor, carelessness, equipment failure, natural disasters and fatal combination of circumstances can lead to an accident with loss of human life.

Generally they concern studying of a state and increase in a safety culture at all levels of management and regulation of nuclear power, the attentive relation to details, implementation of effective programs, identification, the analysis and elimination of the factors reducing safety, and ensuring effective management of knowledge in nuclear area, literacy of personnel. The need of improvement of the quality of training of shots for nuclear power is noted. For achievement of this purpose supplying countries of the NPP have to create the centers of training of specialists for the countries accepting nuclear technologies. A row new is supposed. Issues of non-proliferation of nuclear weapon and prevention of hit of nuclear materials and technologies in hands of terrorists are discussed. The Japanese nuclear crisis will be long and not knowing frontiers.

From 1952 to 2004 in the world on the NPP there were 29 major accidents, including a little from them on the same NPP. According to the data of International Association according to the prevention of influence of radiation (May 2004). From 1950 to 2001 on nuclear power plants about 500 accidents from which about 2 thousand people have suffered are registered. Accidents took place and after the specified information and so far. For the same period of time, there were four accidents on medical installation as a result of which 35 people have suffered. 15 accidents have happened in the industry. 30 people have suffered.

A number of accidents on the NPP vary from a small emission of radioactive materials before heavy accidents.

From 1951 to 1986 in the USA 12 accidents are registered. The first accident has taken place in 1951 in the state Detroit.

In the former Soviet Union from 1957 to 1985, there was information on 12 accidents on nuclear power plants (Belozersk, Leningrad, etc.), including the small Chernobyl accident.

Radiation accidents happen not only on the NPP.

Major radiation accidents very first in the history have happened during an operating time of nuclear materials for the first atomic bombs.

On September 1, 1944, in the USA, the State of Tennessee, in the Obninsk national laboratory in the attempt to clean a pipe in the laboratory device on uranium enrichment there was a uranium hexafluoride explosion that has led to the formation of dangerous substance – hydrofluoric acid. Five people who were at this time in the laboratory have suffered from acid burns and inhalation of the mix of radioactive and acid vapors. Two of them have died, and the others have suffered serious injuries.

In the former Soviet Union, the first severe radiation accident has happened on June 19, 1948, the next day after an exit of the nuclear reactor on an operating time of weapon plutonium on design capacity (an object "A" of Mayik plant in Chelyabinsk region). Insufficient cooling of several uranium blocks was resulted by their local alloy age with surrounding graphite. Within nine days the channel was cleared away by a manual with surrounding graphite, so-called "goat". During an accident, elimination all men's personnel of the reactor and also the soldiers of construction battalions involved to accident elimination has undergone radiation. Insufficient cooling of several uranium blocks was result by their local.

On March 3, 1949, in Chelyabinsk region, as a result of mass dumping by Mayik plant into the Techa River of highly active liquid radioactive waste, about 124 thousand people in 41 settlements have undergone radiation. The greatest dose of radiation was received by 28 100 people living in coastal settlements down the river Techa (an average individual dose – 210). At part of them, cases of chronic radiation sickness have been registered.

On December 12, 1952, in Canada, there was the first-ever serious accident on nuclear power plant. The technical human error has led to overheating and part μ/Sv al fusion of an active zone. Thousands of curies of fission products have got to the external environment, and about 3800 cubic meters are radioactive the polluted water has been dumped directly on the earth, in small trenches near the river of Ottawa.

On November 29, 1955 "the human factor" has led to the accident on the American experimental reactor (the State of Idaho, the USA). In the course of the experiment with plutonium, as a result of incorrect actions of the operator, the reactor has self-destructed, has burned out 40% of his active zone.

On September 29, 1957, there was an accident which has received the name Kitty. In storage of the Mayik radioactive waste in Chelyabinsk region, the capacity containing 20 million curies of radioactivity has blown up. Experts have estimated explosion power at 70-100 tons in a TNT equivalent.

The radioactive cloud from the explosion has passed over the Chelyabinsk, Sverdlovsk and Tyumen regions, having formed a so-called East Ural radioactive trace over 20 thousand Esq. By estimates of experts, during the first hours after the explosion, before evacuation from an industrial site of the plant, X-ray more than five thousand people have undergone single radiation to the 100th. During the period from 1957 to 1959 participated in recovery from the accident from 25 thousand to 30 thousand servicemen. In the Soviet Union accidents were secret.

On October 10, 1957, in Great Britain, there was a major accident on one of two reactors on an operating time of weapon plutonium. Owing to the mistake made at operation, fuel temperature sharply has increased in the reactor, and in an active zone, there was a fire continuing within 4 days. Have received injuries of 150 technological channels that have caused emission of radio nuclides. In total, about 11 tons of uranium has burned down. Radioactive fallout has polluted the extensive areas of England and Ireland; the radioactive cloud has reached Belgium, Denmark, Germany, and Norway.

In April 1967 another radiation incident at the *Mayik* enterprise has occurred .which used for dumping of liquid radioactive waste, the lake Tech. It has strongly shoaled; at the same time, 2-3 hectares of a coastal strip and 2-3 hectares of a bottom of the lake have become bare. As a result of wind rise of ground deposits from the Become bare sites of a bottom of a reservoir radioactive dust about 600 Cu of activities has been taken out. The territory of 1 thousand 800 square kilometers in which about 40 thousand people lived has been polluted.

In 1969 there was a failure of the underground nuclear reactor in *Listens (Switzerland)*. A cave where there was a reactor, infected with radioactive emissions, it was necessary to immure forever. The same year there was an accident in France: on the NPP. The started reactor has exploded. It has turned out that during a night shift the operator by carelessness has incorrectly loaded the fuel channel. As a result, a part of elements have overheated and has melted; about 50 kg of liquid nuclear fuel have flowed out.

On January 18, 1970, there was a radiation accident at the Red Sormovo plant (Nizhniy Novgorod, Soviet Union). At construction of the nuclear submarine, there was not allowed the start of the reactor which has fulfilled at an ultra boundary power about 15 seconds. At the same time, there was the radioactive infection of a zone of the shop in which the vessel was under construction. In the shop, there were about 1000 workers. Radioactive infection of the area managed to be avoided because of the closeness of the shop. That day many have left home, without having received necessary decontamination processing and medical care. Six victims were taken to the Moscow hospital; three from them have died in a week of sharp radiation sickness. From the others hand have taken a subscription about nondisclosure of the event for 25 years.

The main works on the elimination of accident were continued until April 24, 1970. More than one thousand people have taken part in them. By January 2005 from them, 380 people have survived.

The seven-hour fire on the reactor of the NPP (state of Alabama, USA) has cost on March 22, 1975, 10 million dollars. Everything happened after the worker with the lit candle in a hand has got to close up an air leakage in a concrete wall. The fire has been picked up by a draft and has spread through the cable channel. The NPP for a year has been put out of action.

The accident on the NPP Tree mail Island in the State of Pennsylvania which has happened on March 28, 1979, became the most serious incident in a nuclear power of the USA.

In the night of April 26, 1986, on the fourth block of the Chernobyl NPP (Ukraine), there was the largest nuclear accident in the world, to the partial destruction of an active zone of the reactor and an exit of splinters of division out of zone limits. According to experts, the accident has arisen because of the attempt to do an experiment on the removal of additional energy in operating time of the main nuclear reactor.

On September 30, 1999, there was the largest accident in the history of the nuclear power of Japan. At the plant on a production of fuel for the NPP in the scientific town of Tokaymura (the prefecture of Ibaraki) because of a human error uncontrollable chain reaction which continued within 17 hours has begun. 439 people have undergone radiation, 119 of them have received the dose exceeding annually admissible level. Three workers critical doses of radiation have received. Two of them have died.

On August 9, 2004, there was an accident on the NPP located in 320 kilometers to the west of Tokyo. In the turbine of the third reactor, there was a powerful emission of steam with the temperature about 200 degrees Celsius. The staff of the NPP who was nearby has got serious burns. At the time of an accident in the building where the third reactor is located, there were about 200 people. The leak of radioactive materials as a result of the accident isn't revealed. Four persons have died, 18 – have seriously suffered. The accident became the most serious a number of the victims on the NPP in Japan.

On March 11, 2011, in Japan, there was the most powerful for all history of the country an earthquake. As a result on the NPP, the turbine has been destroyed; there was a fire which managed to be liquidated quickly. On Fukushima Plant-1 the situation has developed very seriously - as a result of the shutdown of the cooling system nuclear fuel in the block nomber1.The reactor has melted, outside of the block radiation leak has been recorded, in a 10-kilometer zone around the NPP, evacuation is carried out. Next day, on March 12 mass media have reported explosion on the NPP.

Thus, in 67 years (1944-2011) in the world fix14 various weight of atomic accidents.

These data are regularly replenished. At the experts, the weight of evidence suggests that the specified figures are underestimated and continue to be underestimated.

The scale of integrated assessment of the risk of major radiation accidents on the NPP is offered. According to the specified scale, the Chernobyl accident is estimated at seven points, at Three Mile Island – five points, in Chelyabinsk – six points, in Tomsk - four points. Test of atomic weapons is above scale indicators.

Along with big accidents on nuclear power plants, limited malfunctions which victims, first of all, are service personnel are possible. In the former Soviet Union similar information kept under "seven seals".

After the accident on the American nuclear power plant in Tree mail Island (Midtown, the State of Pennsylvania, 1979), their construction in the USA has been suspended, despite the lack of the official ban. Recently this question is intensively discussed and, apparently, construction of the NPP will gain further development. It has been confirmed in the inauguration speech of the U.S. President J. Bush (2005). Time will show further development of nuclear power. The importance of nuclear power is maintained by the President Barrack Obama.

It is necessary to emphasize that the danger of nuclear power plants is connected not only with a possibility of accidents, and to other reasons.

First, on many of them, atomic arms are confidentially made. Secondly, it is impossible to exclude a possibility of plunder of radioactive materials for terrorism.

The specified questions are in the center of a research and attention of large experts of the USA and some other countries.

The further succession of events has shown that all drama not in discoveries, and in them not always expedient and correct use. There is a set of the Regional, National and International bodies exercising control of a distance of the NPP and developing actions for diagnostics, prevention, and treatment of radioactive effects, in particular, on the person

The energy crisis which is outlined in the world forces many countries to increase nuclear power (Ecological organ Greenpeace, MAGATE, Doctors without Borders and others).

The special danger is constituted by the so-called floating nuclear power plants (FNPP) developed in Russia. Their introduction in some southeast countries is begun. According to experts of PAES represent big danger since can be widely used by terrorists.

It is impossible to tell about all accidents on the NPP. We have chosen the cases which are most fully lit in literature. And meanwhile, along with the community, accidents in many respects differ from each other in a number of features.

Accident on the American nuclear power plant Tree mail Island

It has occurred on March 28, 1979, on the second reactor of nuclear power plant. During cleaning of the reactor, there was an accident as a result of which there has occurred insignificant leak of radioactive materials in the atmosphere. The specified accident is the heaviest in the history of the nuclear power of the USA.

The temperature in the reactor during accident reached 2200 degrees; about a half of all components of an active zone has as a result melted. In absolute figures, it makes nearly 62 tons. From the nuclear reactor a large amount of radioactive water, therefore, radioactivity level in rooms of containment more than by 600 times has exceeded norm has followed. The number of radioactive gases and steam has got to the atmosphere, and as a result, each inhabitant of a 16-kilometer zone around the NPP has received radiation no more, than during the fluorography session. The most dangerous — emissions in the atmosphere and water of highly active nuclides managed to be avoided therefore the area remained "clean".

At once after an accident the wide complex of actions including is held:

- definition of a collective dose of radiation;
- forecasting of number of cases of additional diseases of cancer among lived inhabitants in a radius of 50 of a kilometer zone;
- forecasting of a number of additional cases of developing of other fatal and serious illness.

The damaged reactor was immediately turned off and in a short time repaired. Started intensive cleaning of the contaminated area. And only in 1993, the area was cleaned of radioactive substances. Soon received the permission to the operation of the refurbished reactor. Currently, the damaged reactor is periodically activated.

Monitoring of affected persons and survey of the contaminated territory continues. After the accident the Governor of the state, in consultation with eminent experts, recommended that pregnant women and preschool children living within a radius of 50 kilometers from the reactor will all move into the pure regions.

All recommendations were supported by the organizational actions and the full appropriation. What is happening truthfully and promptly reported in the media (TV, radio, Newspapers, etc.). And now for the past almost 30 years, all employees of a nuclear power plant (1000 people) and persons living in the 50-kilometer zone are under the regular and multidisciplinary supervision. Examination of this group of people, aimed at identifying the most frequent effects of radiation – cancer of different localization and deformities of development.

NPP Tree mail Island

Comparatively the small troop landing of radioactive materials in combination with the operatively conducted prophylactic, curatively-diagnostic and ecological events was prevented by human victims and heavy consequences.

For indicated a period of time, independent researchers are expected doses irradiations that the looked after the group of persons could get. She averaged 100 μ/Rem, that corresponds to the 1/3 natural radiation in the USA.

Painstaking observations haven't found a negative influence of accident on the health of the persons which have undergone radiation. Part of observed persons had had unshapely expressed fear of radiation, i.e. fear in communication by a possibility of negative influence of radiation on health

Nevertheless, observations of the specified group of persons including their posterities proceed.

Comparison of non-selective radioactive material in software with surgically performed prophylactic, medical-diagnostic and environmental activities that prevent human casualties and these are the consequences.

Avalanche did not have any serious consequences for the welfare of scientists and scientists, but it had only a significant impact on the United States and the Americas. But, do not go to work, but do not worry about it for 1993 years.

At present, AMS has been operating - a functioning power plant No. 1, which has been temporarily degraded, and has been depopulated in 1985. Secondly, the reactor has been used and used, and for square observation. The station is working until 2034.

After the accident on the NPP Three Mile Island in the USA the decision not to build nuclear power plants anymore that has resulted in stagnation in the American nuclear power has been made.

The psychology of people has changed. Have appeared, the so-called "Chinese syndrome". On the surprising combination of circumstances in two weeks prior to the accident to big screens, there was a movie "Chinese Syndrome" narrating about the accident on the NPP. The slangy term "Chinese syndrome" which is thought up in the 1960th years by nuclear physicists means accident at which fuel in the reactor melts and burns a protective cover. And in the second power unit of the NPP Three Mile Island, there was a reactor meltdown! So there is nothing strange that after real accident panic has risen, and no assurances of the high-ranking officials, including the U.S. President, could calm people finally.

Accident on Chernobyl nuclear power plant in Ukraine

Construction of the Chernobyl Nuclear Power Plant (CNPP) that in 110 km from the capital of Ukraine of Kiev dropped out for 60-70 of the 20th century. During this period in the republics of the former Soviet Union, there was a boom of the wide use of so-called peaceful atom, in particular, of nuclear power. The made decision on construction of the CNPP, despite shy objections of ecologists and other experts, had to be surely executed in the put terms.

To a huge regret, constructive shortcomings and defects of the fourth reactor haven't affected the terms of his start appointed by Government agencies and party face.

For the small period of time, six nuclear reactors are constructed. The name of V.I. Lenin is hastily appropriated to the new NPP.

From recently opened archives it became known that knew about a possibility of the accident in two months prior to her emergence. This information was secret therefore no preventive actions were taken.

On the night of April 26, 1986, during a planned inspection of systems of the fourth reactor et the Chernobyl nuclear power plant there was an explosion.

The sharp power surge lasting for only 20 seconds was his reason. As a result of the explosion the roof is broken, the fourth reactor is completely destroyed and radio nuclides have received a free exit in the atmosphere. After the opening of archives also other sad facts which have led to accident became known. Their contents have no fundamental differences.

Thus, the technical defects multiplied by a human factor have caused the heaviest accident in the world on nuclear power plants.

Later the Nobel Prize laureate Hans Bets has told: "The project in the basis is illegal".

Studying of causes of the accident and their assessment continue some of her mechanisms and consequences are specified.

Earthquakes are specified in classifications of the reasons of accidents on the NPP. Fortunately, almost 60th summer history of nuclear power didn't know accident cases on the NPP caused by an earthquake. Information which has appeared in 9-10 years after the Chernobyl accident was an exception. Authors claimed that the local earthquake in the location the fourth reactors were a cause of the accident. These data haven't found confirmation and support. After the accident on the Japanese nuclear power plant information has renewed, but in the subsequent has stopped. As a rule, the NPP build in not seismic countries. They are calculated on an earthquake up to 5-7 magnitudes.

An accident in 1988 on the Matsemorsky NPP located near epicenter (Spitak, Armenia) was an exception. As a result of an earthquake measuring about 7 with aftershocks till 9-10 magnitudes have stopped the reactor for six and a half hours then some time worked with interruptions, haven't been disconnected finally yet.

The leading staff of the NPP considers that small accident at a strong earthquake can be connected with the high quality of the equipment. Besides, in Japan, there were two factors – an earthquake and the strongest tsunami. Seven years ago the representative of the European Union called an object a source of danger to all regions. The requirement of immediate construction of the new station was made also by the U.S. Government. A new object in the same place is planned to be put into operation after 2016.

Accident on CNPP has happened on Friday night. There was in the afternoon fine spring weather and residents of Pripyat in which the CNPP is located, and vicinities have gone to picnics, fishing, etc. Children played in gardens and parks. Life proceeded in the usual course the day of rest.

Only the small group of experts has been called for work in connection with the accident happened earlier. Therefore it wasn't alarming. Evacuation of the population of the zone, next to the reactor (in a radius of 30 km) has begun only in 36 hours after the explosion.

The first official information on accident has appeared for the third day after the incident. The short message in extremely optimistic tone is published in the Republican newspaper "Soviet Ukraine ". So limited information was in the all-Union newspaper. The head of the government M.S. Gorbachev who was on vacation in the Crimea knew nothing at all (?!). He has for the first time acted on TV on May 15, 1986.

For comparison, there is a wish to note that at once after the accident to the NPP in Tree mail Island there has arrived the U.S. President J. Carter. Also, the head of the government of Japan after the accident on the NPP in 1999 has arrived.

Big unjustified criminal optimism characterized also the subsequent information of Allied and Republican Government agencies. The people were assured that the situation is under control. There is no reason to panic.

The fourth reactor of the CNPP before and after accident

Nobody did give any specific instructions on places and the nature of radioactive infection (the soil, plants, water, etc.).

To the Government message, the staff of Institute of physics of Academy of Sciences of Ukraine who was carrying out the planned inspection of a radiation background in Kiev region has noted his sharp increase. Having consulted to the management, have estimated the found changes as result of the malfunction of dissymmetric devices.

On April 28, 1986, in separate medical institutions of the city of Kiev, including the institute at which I worked, order about the preparation of places for injured with the accident is directed to the CNPP. However, no concrete recommendations of an essence of preparation existed.

In the same time inhabitants of Kiev region and the city of Kiev have paid attention to a large number of the buses filled with people, moving from the West to the east. A part of buses approached city baths where "processing" of victims was carried out (a shower without change of linen and footwear). The bath located in the populous place of the center of Kiev opposite to Republican stadium was one of such points. The polluted buses stood idle on the street till many hours. People dispersed who where could (remained with the family and acquaintances, left, etc.) and spread radioactive materials with clothes and footwear. Distribution of radiation was also promoted by numerous transports from the polluted zones.

In a zone of accident representatives of service of radiation control, civil defense forces chemical, with the Ministry of Defense, and the Ministry of Health worked.

Besides accident elimination, their task included measurement of a radiation situation in the NPP and a research of radioactive pollution of environments, evacuation of the population, protection of an exclusion zone which has been established after the accident.

Literally from the first day o the accident rising of the level of background the radiation was registered in the Scandinavian countries, Poland, Czechoslovakia, Austria, the southern Germany, and northern Italy. Then, in connection with the change of the direction of the wind, the radioactive cloud appeared over the Balkans, Greece, Turkey and two states of the USA. Foreign radio stations ("enemy voices") reported about it. It wasn't recommended to listen to them to the Soviet citizens categorically. Nevertheless, this information gradually became the property of some segments of the population.

Along with the specified criminal lie, higher party and Soviet persons in Moscow, Ukraine, and Belarus were informed on the difficult character of the accident, some methods of personal prophylaxis and need of evacuation of children aged up to 14 years. The specified information went under a signature stamp: "Top secret". The people of the country, including affected regions, continued to a irritate, call criminal for a normal way of life, participation in a May Day demonstration, a parade and picnics on the occasion of the celebration of a Victory Day (on May 9), fishing. It was very bitter and heavy to perceive this entire lie.

What has really happened?

As a result of the explosion et the fourth reactor from 100 to 150 million has come to the atmosphere. Curie. The specified figure in many tens times more what took place at the explosion of atomic bombs in the Japanese cities of Hiroshima and Nagasaki.

Radioactive materials were in three physical shapes – gas, aerosols and firm fragments.

The unprecedented scale of an accident is connected with two reasons:

1) explosive depressurization of an active zone of the reactor;

2) burning of the graphite laying caused by the repeated explosion which has come in a few minutes.

All this has increased the volume and duration of emission of radioactive materials (over 15 days).

The radioactive range included over 200 names of nuclides.

The most powerful emission of radioactive products was observed in the first two - three days after the accident.

For the sixth days, the power of emission began to grow because of a warming up of an active zone up to the temperature over 2000 degrees Celsius. By ninth days the power of emission has grown to 60% of initial. After cooling of an emergency zone there has occurred reduction of power of the explosion. However, then two periods of increase in the power of the explosion have followed. It was among May 8-11 and, especially on May 14-17, with a maximum on May 16, 1986.

From quite recently the opened documents it became known that preliminary conclusions have been drawn on the weight of accident by May 4, and final by May 11, 1986. Unfortunately, it hasn't found.

Among radioisotopes at the beginning of accident radioactive iodine prevailed (iodine - 131). Further, the range of radioisotopes has sharply extended (caesium-141,144; ruthenium-103, zirconium-95, strontium - 89 and many others).

It should be noted that a certain number of isotopes of radioisotopes changed and continues to change in various terms after the accident. So, rather recently in a zone of accident very aggressive isotope americium is found. He is characterized by a huge half-life period and high extent of penetration. It is predicted that in 50 years americium will figure prominently most in a range of radio nuclides.

Along with isotopes the huge amount of chemicals, a part of which (for example, lead) were used for fire extinguishing on the fourth reactor, has got to air. As it has appeared later, use of lead had no reasons practical use. Because of an open active zone, the released radioactive materials quickly rose highly up (1500 km), forming a radioactive cloud. Atmospheric streams carried a cloud in compliance with a force and the direction of the wind. Due to the rains on the course of advance of the specified cloud, there was a big loss of radioactive materials in the atmosphere.

Wind radioactive materials have spread to Ukraine, Belarus and certain districts of four western regions of Russia. There are informal data that by means of special devices the direction of the wind has been changed to Moscow to the west.

In Ukraine, 12 areas have undergone radio isotope pollution: Chernihiv, Cherkassk, Chernovitskaya, Ivano-Frankivsk, Kiev, Kirovohrad, Rivne, Sumy, Ternopil, Vinnytsia, Volynsk, and Zhytomyr. The highest dose of radiation was received by the population of the Zhytomyr, Kiev and Rivne regions.

According to the data of the Ministry of Health of Ukraine, in the infected territories over 17 million people lived.

More than 4.6 million hectares of fertile lands are struck. 70% of radiations were the share of Belarus.

Radiation pollution has captured five areas of Belarus: Brest, Gomel, Grodno, Mogilev, and Vitebsk.

In Russia, radiation pollution is registered in certain districts of the Bryansk, Kaluga, Oryol, and Tula regions. In Russia, the area polluted by radioactive materials was more than 59 thousand

Esq., including agricultural grounds of 2.9 thousand hectares and about one million hectares soil forest fund.

In Belarus from the direct influence of radiation more than six thousand Esq. of lands, including about three thousand Esq. fertile are brought out of economic use. The population of the infected areas made over two million people. Villages have become empty, some kindergartens and schools were closed only in 1990. The Supreme Council of the Republic all territory has been announced by a zone of ecological catastrophe.

About 60% of radiation exposure has fallen to a lot of country people. The total area of infection with radioactive materials, according to Ministry of Atomic Energy of the former Soviet Union, has exceeded 130 thousand Esq. with the population of 20 million people.

In these territories, nearly 1 million 800 thousand people continue to live. From the polluted territories, 52 thousand people have been moved in an organized order, or independently. Is sad the fact that over 30 years later after the accident, places with the high level of pollution in which people continue to live come to light.

By the last estimates of various experts five - seven million people live in conditions of radioactive pollution so far.

It became known later that radioactive fallout has dropped out in the Leningrad Region, Moldova, and Chuvashia. Subsequently, pollution has been noted in the Arctic regions of the USSR, Norway, Finland, and Sweden.

It is necessary to emphasize that data on scales of radiation defeat (force of the explosion, the areas of contamination, the number of victims, etc.) as a result of the accident on the CNPP differ in certain fluctuations. The specified circumstance is caused by a number of the subjective and objective reasons.

It is necessary to refer incompetence, falsity, distortion, and concealment of the true facts by the government organizations of the Soviet Union to the subjective reasons and under control bodies then still of the Ukrainian, Belarusian, and Russian republics. The disorganization, incompetence, and confusion were looked through in everything.

Unauthenticated of information remains also in independent Ukraine, Belarus, and Russia.

The objective reasons include poor quality and the insufficient number of dosimeters, insufficient preparation of technical and medical shots and also bad material resources. And, at last, the variability of data can be explained, to some extent, with the distinction of a half-life period of a range radioactive SUBSTANCES. Some value has the natural and artificial shift (large construction, spontaneous influences, etc.) of terrestrial layers.

The radiation contamination caused by accident on the CNPP has a number of features. Treat them: spottiness (unevenness) of pollution.

The various structures of nuclides in many respects depending on the distance to the damaged reactor; various mechanisms of distribution of nuclides in a chain the soil - plants people.

Taking into account influence on the person, four periods (phase) of the Chernobyl accident are allocated.

 1. Period super early – formation of a radioactive cloud in the course of which there are a radiation very high doses getting to an organism, mainly, through respiratory organs. The main radiation is the share of lungs, thyroid gland, and digestive tract.

 2. Period the iodine, lasting 1-2 months during which there was all load thyroid gland.

In the subsequent time, it became known that true doses of radioactive iodine haven't been precisely defined because of misuse of dosimeters.

 3. Period lasting two-three years are characterized by external gamma irradiation by generally short-lived radionuclide (barium - 140, 141, etc.). At this time the main part of internal radiation at the expense of caesium-134, 137, coming to a human body with food and water are implemented.

 4. Period -the influence of cesium, strontium plutonium, lasting several years to tens and hundreds of years (Table 1).

Many classified documents became available to studying only 15 years later after the accident. Time will show as far as they are fully presented today. Specification of the sizes and consequences of the Chernobyl accident continues and will continue for a long time.

The constant lack of accurate information led to the emergence of various rumors, including the emergence of unskilled recommendations. So, for example, for prevention and treatment of radiation defeats alcohol was recommended. It isn't difficult to imagine what support of the population the recommendation has found. Events for clarification of places of pollution weren't less with gross spelling mistakes held. Houses poured down from fire pumps, and water flew down on the surface of streets, gardens, etc. Ladder flights in houses without any sequence plentifully poured down water. Leaves gathered (generally students) and then were burned in the inhabited places. Practically there was no radiological control of foodstuff, as in public institutions, and the private sector. The list of a similar outrage can be continued.

Single scientific practical publications (so-called methodical letters and newsletters) were inaccessible to the population. In avoidance of panic (?!), information went to elite institutions and departments. She could be are used only by physicians and that with big restrictions (special permissions, work in certain time, etc. were required).

The Soviet functionaries of all levels, with the assistance of KGB, without troubling itself the clarification of true causes of the accident, have hurried to declare that she is connected with negligence of officials of the CNPP.

At once after the accident on the fourth block of the CNPP, the work directed to the termination of the leak of radioactive materials has begun. People, risking life, without special knowledge and the equipment, did everything possible also impossible for reduction of consequences of the accident. The only being available "Robot" has failed soon. The Soviet government refused the "Robots" offered by other countries. To replace the broken "Robot" soldiers are sent. Without special knowledge and methods of protection, they have continued work. Humorists with bitterness called their "Bio robot".

Soon after the accident the special shelter, a so-called sarcophagus is built.

However, the leak of radioactive materials proceeded and continues. Several years later became obvious that the specified construction doesn't ensure safety, as in the present and future time. The government of Ukraine had to give way to ambitions and to ask the help in the developed countries and the authoritative international organizations (WHO, the UN, and many others).

The world, state and local organizations repeatedly made decisions on the closing of the destroyed reactor. The considerable value was rendered by various public organizations and the movements around the world, in Ukraine, Belarus, and Russia. At last, only at the end of 2000 under pressure and with big material support the *President of Ukraine Leonid Kuchma* has given the order on a closing of the fourth reactor of the CNPP. This time was repeatedly postponed.

For a number of the years, works directed to an improvement of a sarcophagus were continued. And only in November 2008 signing of the act of his reception has taken place. How today projects of the shelter of the fourth reactor are again discussed. At the initiative of the Commission of the European Community and the Government of Ukraine, it is developed and the Shelter Implementation Plan-SIP project has begun to be carried out.

The specified project assumes transformation of the existing sarcophagus to the safe and ecologically stable system called a confinement. The design term of his service is-100 years. The shelter has to provide also the safe dismantling of old designs and extraction fuel-bearing, in other words – radioactive, materials qualitatively and honestly it is carried out will show the future.

A number of the countries take part in this project. "Shelter" represents an arch 108 meters high, 150 meters wide and flight of 257 meters. For descriptive reasons it is possible to tell that under her American "Statue of Liberty" and Leningrad Motherland can be located.

Now the large protective construction "A new safe confinement, – an arch which will be pulled from "Shelter" is built.

Shelter ("sarcophagus") of the fourth reactor

It is necessary to emphasize that after closing of the CNPP the danger of radiation pollution will remain. It is connected with possible floods, dumping of the polluted waters, wildfires of the trees damaged by radiation. Besides, even the closed CNPP conceals in itself danger and is capable of new surprises. In the nuclear fuel which has remained in the destroyed power unit, uncontrollable chain reactions begin, according to scientists. There was an americium which is one hundred times more dangerous than strontium.

In the site of the Chernobyl NPP construction of dry storage for the fulfilled nuclear fuel (continues. Now, the fulfilled fuel on the platform of the CNPP is stored in the temporary storage of "wet type" constructed in the Soviet period and in cooling ponds. The temporary storage of spent nuclear fuel will provide a possibility of "dry" storage throughout the period not less than 100 years of more than 20 thousand fulfilled heat allocating assemblies. The project is financed from the Account of nuclear safety and kept by the company Toltec International, USA.

The certain exaggeration of radiological the consequences of the Chernobyl accident which is warmed up by unfair mass media has generated in the consciousness of the victims a condition of hopelessness and hopelessness that are the reason of a stressful state. An unambiguous conclusion which was drawn by the population from the tragedy: in case of similar the accident on the NPP the person loses all – health, and the children and relatives, work and property.

For a considerable part of the population any radiation – the reason of various diseases, genetic disorders, fatal ontological diseases. Such perception – the steady and reproduced in new generation's phenomenon of mass consciousness – has caused a crisis of confidence in nuclear power.

Inadequate perception of radiation hazard, according to experts, has the objective concrete historical reasons among which:

• concealing by the state of the reasons and real consequences of the accident;

• ignorance by the population of elementary fundamentals of physics of the processes happening both in the field of nuclear power and in the field of radiation and radioactive influence;

• hysteria provoked by the mentioned reasons in media;

• numerous problems of the social character of all-federal scale which have become the good soil for the fast formation of myths and so forth.

Influence of accidents on the health of the person depends not only on its power and a nuclide range. Timely, competent, honest carrying out a complex of organizational and preventive actions, including medical is of very great importance. It is brightly illustrated on the example of the Chernobyl accident. The fact that there was on the CNPP not only an accident, and heavy accident. She treats the biggest technogenic accidents, become the largest ecological disaster of the left century. The conclusion is absolutely fair that accident on the CNPP was also the moral category.

Sᴛatue Prometheus

After Ukraine has found independence, the law forbidding to code data on environmental disasters. However, has been adopted it has a little helped to open of information on the Chernobyl accident. In 2006, in 20 years after the Chernobyl accident, the Sᴛatue Prometheus has removed a signature stamp "confidentially" from a number of the documents connected with the accident. However, completely KGB-People's Commissariat for Internal Affairs archives for 1917-1991 including those that concern the CNPP have been open. It has occurred thanks to the law on access to archives of repressive persons of the communist totalitarian regime of 1917-1991 adopted a year ago in April 2015.

For last 30 years in 122 cases of leukemia among liquidators are recorded. 37 of them could be induced by the Chernobyl radiation. Increase in quantity of diseases of other types of oncology among liquidators in comparison with other groups of the population hasn't been recorded.

During the period from 1986 to 2011 from 195 thousand Russian liquidators registered about 40 thousand people have died of the different reasons, at the same time the general indicators of mortality didn't exceed the corresponding average values of the population of the Russian Federation.

It is established that for the end of 2015, from 993 cases of diseases of cancer of a thyroid gland at children and teenagers (at the time of the accident) 99 could be connected with radiation exposure.

Listed has to be an attention subject as in affected countries, around the world. Prominent scientists, the authoritative world organizations (the UN, WHO, etc.) call various communities and the movements for it.

Everything depends on what consequences and obligations are involved by recognition of this or that fact for the concrete organization or the state. It is probable therefore reports on a research of consequences of the Chernobyl accident time looked odious at all. So, for example, the report of the scientific committee on the action of atomic radiation of the UN "Human consequences of a nuclear incident in Chernobyl" made in 15 years after the accident makes an impression that anything special hasn't occurred. Main conclusions of the UN of 2001: a) impact of radiation on human health was smaller, than assumed fifteen years ago; b) more harm was done by resettlement of people from affected areas: families, unemployment among the compelled migrants, the depressions and diseases caused by a stress are destroyed; c) privileges, holidays, food and medical care which is received by the victims of Chernobyl have made them dependent on the state and have cultivated in them feeling of fatalism and pessimism; d) still there is no International recognized evidence of increase in frequency of a disease of a leucosis among the population living in the territories polluted by radio nuclides and also among those who worked at restoration of nuclear power plant. There is no statistically significant evidence of growth and other cancer diseases and also the birth of children freaks. The most part of photos of babies with congenital defects which have been used by the western charitable organizations for collecting donations, actually a photo of those which diseases have nothing in common with Chernobyl.

The reasons of concealment of the true situation in the USSR right after the accident and in the West or in the same Ukraine years later different. At first, the authorities in the Soviet Union tried not to allow panic, but, nevertheless, organized the evacuation of the population. In Ukraine, 9.1 thousand people have been evacuated from them about 50 thousand – inhabitants Pripyat. And all from an exclusion zone, about 200 thousand people in three former Soviet republics have been evacuated. Seriously polluted there were territories of Ukraine, Belarus, and Russia with the population approximately in 7. 1 million people with a total area of 155 thousand Esq. Now concealment of consequences in Ukraine and misrepresentation is the result of impossibility and unwillingness of the "independent" state to fulfill the obligations to the population in the conditions of the wildest capitalism which is rolling down in neofeudalism.

In the in subsequent years in memory of events and people of the Chernobyl accident monuments and memorials are constructed, movies, works of art are created. In Kiev the museum is open.

Accident on nuclear power plant in Japan

On September 30, 1999, on the nuclear power plant Tokaymura which is 120 km to the northeast of Tokyo, there was an accident. It was caused by a mistake of two workers at the time of loading of the reactor radioactive fuel.

Indoors there were three more persons. Two of them have died from sharp radiation sickness the Dissymmetry has shown that the received dose varied from 24.5 to 3.9 Gr. In total in the territory of the station, there were 229 employees in hum the dose of radiation has made 0.07-0.48 µ/Sv.

Nuclear power plant in Japan

Combination of the phenomena, extremely improbable on the scales (historically maximum earthquake measuring 9 magnitudes with, the followed historically maximum tsunami), have led to a loss of all power supply at the station. In combination with the subsequent the earthquakes, the cooling system has failed. On several power units of the NPP, there were explosions that have led to an emission of radioactive elements in the atmosphere. From six power units, four have been damaged. In the first days after the accident in superheated reactors and pools with fulfilled the heated fuel cores filled tens of thousands of tons of sea water. Partially then she has got to basements and has filled in the drainage system of the reactor. Very infected water from the drainage system of the second power unit through a breach poured out to the ocean that has brought in a zone, adjacent to the station, to sharp jump of radiation level.

NPP Tokaymura after accident

Liquidators could close up a hole; however, the threat of new leak remains. The technology of works on the elimination of accident with the participation of other countries as necessary changes, but is far from the end. So, as of April 26, 2011, emission of radiation made 900 µ / Sv in an hour. The specified indicators change and will change.

Various degrees of severity of acute radiation sickness have arisen at 229 people, including beam burns – at seven. Three patients have died. At 226 patients recovery from ARS has occurred in terms of 82 up to 210 days. 207 people lived in the radius of 350 meters from the reactor. The dissymmetry has shown that the received dose of radiation made 01-0.21 µ/Sv. All employees and inhabitants of the next zone have been urgently evacuated to safe places. To the persons living in the radius up to 10 km recommended not to leave before evacuation rooms. Schools and shops were closed. The accident has quickly received a response of the international organizations and institutions, various communities.

Despite intensive and timely treatment, five people who were near the reactor have died of sharp radiation sickness and its complications.

Lessons of the accident have been carefully analyzed, accepted to execution and published.

Later the presented accident consequences on a nuclear object of Tokaymura differ from preliminary a little. Three workers who were directly working with solution were strongly irradiated, having received doses: one from 10 to 20 Sievert, other from 6 to 10 Sievert, the third from 1 to 5 Sv (while in 50% of cases the dose about 3-5 Sievert is deadly). The first has died in 12 weeks, the second - in 7 months. In total 667 people have undergone radiation (including employees of the plant, firefighters, and rescuers and also locals). Except for three workers mentioned above, their doses of radiation were insignificant (no more than 50 µ / Sv.

The 4th level on the International scale of nuclear events (INES) has been appropriated to this incident. According to conclusions of IAEA, the cause of the incident the human mistake and serious neglect as the principles of safety" have served ".

Further researches and detailed information on results proceeds. Below information for summer of 2016 is provided.

It is established that the place for radioactive water in the radioactive wood at the NPP will end soon it wasn't succeeded to put out. The radioactive wood has lit up. Three Regions of Fukushima Prefecture have completely cleaned from radiation. Near the NPP have opened shopping center. Three thousand residents of Fukushima Prefecture have left the houses. Three new earthquakes measuring up to 5.5 are recorded the tsunami threat is announced. The dismantling of a protective leak of radioactive water on the emergency NPP is found in the emergency. The first reactor of the Japanese nuclear power plant is stopped for checks. The third reactor of the NPP is put into operation.

In underground trenches of Fukushima, more than 10 thousand tons have accumulated. Inhabitants of the next city to Fukushima-1 will return to the houses.

Ground from Fukushima will let on construction of roads and dams in

In Japan, fund raising for victims on Americans is begun.
The presented materials demonstrate the diversity of consequences of the accident.

Accident on the NPP in Japan

On March 11, 2011, the earthquake measuring 9, 0 has shaken Japan and has caused a tsunami which has fallen upon east coast of the country, destroying houses and communications, claiming the lives of hundreds of thousands of people.
This accident became the largest since Chernobyl and has shown, security systems on the Japanese nuclear power plants are now vulnerable.

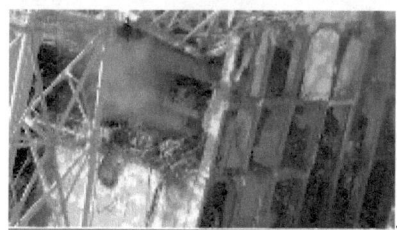

Fusion of the reactor of the NPP Fukushima

The seventh reactor has been appropriated to nuclear accident – the highest level on the international scale of nuclear events). By calculations of the Agency of the nuclear and industrial safety of Japan (Nuclear and Industrial Safety Agency – NISA), the amount of the radioactive caesium-137 released into the atmosphere during the accident is comparable to 168 bombs dropped on Hiroshima in 1945. About a possibility of fusion of fuel owing to loss of the cooling system, and that it can happen as a result of tsunami blow in 2008 it was said in the documents published by the Organization for the nuclear safety of Japan (Japan Nuclear Energy Safety Organization). The owner of the NPP – the Tokyo energy company) knew that Fukushima won't bear a brunt of the disaster. However, the company hasn't made anything to increase the safety of work of the station and has simply ignored the possible danger.
The earthquake became the reason of the failure of power supply on the NPP with six atomic power units. The tsunami has flooded reserve diesel generators, and the station was left without power supply which is necessary for the operation of the cooling system of reactors. As a result, nuclear fuel 1, 2 and 3 of reactors has begun to melt. Because of hydrogen congestion in buildings where reactors are located, destructive explosions have taken place.
From above 150 thousand people have left the infected territories in a radius of 50 km from the NPP. In a 20-kilometer evacuation zone entrance as experts consider that these lands are unsuitable for life within the next decades is still closed. Most of the people evacuated from more remote territories will hardly decide to return to former places – they are afraid of radiation, unemployment and don't want to live in "ghost towns".
By results of a research which was as conducted by scientists from Oceanographic society, the Fukushima accident became the reason of "the largest emission of radiation for all history to the World Ocean". In April 2011 in the samples of ocean water taken at the coast of Fukushima, the caesium-137 level of 50 million times exceeded level before the accident.
According to scientists, within the next decades, it is impossible to foretell how radiation will affect ecosystems. In samples of seaweed and fishes, the content of radio nuclides strongly exceeds maximum permissible norms. If radionuclide gets to a food chain, then strontium which is capable to collect in a human body can increase the risk of diseases of leukemia and cancer of bones.

Trace radiation has been found in rice, meat, fruit, vegetables, milk and baby food. All this causes attacks of fear and panic in the population and a heavy burden lay down on the Japanese economy. In January 2012 the Ministry of Economics, trade and the industry of Japan (Ministry of Economy, Trade, and Industry) admitted that radioactive gravel was used for the construction of new houses, repair of roads and other infrastructure injured with an earthquake. Any rules for radiation control of building materials (a stone, gravel), it was accepted not.

It should be noted a certain ambiguity of information of the certain countries and departments throughout all post - emergency period. According to the available data (October 2011) leak of radioactive materials remains. At the same time, their reasons aren't always clear. Complete elimination of consequences of the accident will require not less than 30 years. At the moment it isn't possible to carry out the comprehensive analysis of this tragic event due to the lack of exhaustive information. The forecast is submitted not less difficult. Radiation from the accident on Fukushima makes 10% of that in Chernobyl. Any person after the accident on Fukushima hasn't got sick with ORS. It is caused by rather low radiation level and the regulated shift. According to the most conservative emergency regulations developed after the accident of a dose of radiation have to be within 150 μ/ Sv but it is admissible also to 250 μ/ Stars - taking into account that even triple radiation won't lead to sharp radiation sickness. At the same time, it should be noted that calculation of the doses received during the emergency works, very difficult task. All participants of work on elimination of the accident are under constant observation. During the post emergency period CHANGED the radius of resettlement of the population - from 20 to 40 km. Settlements for resettlement are added. 970 children, for the purpose of prevention of damage to the thyroid gland, receive iodide potassium. By analogy with other heavy radiation accidents, it is possible to believe that her consequences will be shown in 10-20 years.

Accident on the NPP wasn't an exception in the emergence of the false information provided by a little competent or incompetent people. Similar information promotes panic and unjustified use of some medicines. The same treats methods of prevention and treatment of radioactive effects and their forecast. So, the Japanese police have detained two distributors of drugs "from radiation", about 600 thousand dollars which have managed to earn on medicine sale. It is undoubted that further studying of various aspects of the tragedy will reveal a number of features of diagnostics, treatment, and prevention at the victims from radiation. They will have to be brought to the population.

More than 150 thousand people have left the infected territories in a radius of 50 km from Fukushima Plant. In a 20-kilometer evacuation zone entrance as experts consider that these lands are unsuitable for life within the next decades is still closed. Most of the people evacuated from more remote territories will hardly decide to return to former places – they are afraid of radiation, unemployment and don't want to live in "ghost towns".

By results of a research which were conducted by scientists from Oceanographic society, the Fukushima accident became "the largest emission of radiation for all history to the World Ocean". In April 2011 in the samples of ocean water taken at the coast of Fukushima, the caesium-137 level of 50 million times exceeded level before the accident. According to scientists, within the next decades, it is impossible to foretell how radiation will affect ecosystems. In samples of the seaweed and fishes taken for tests by experts, the content of radio nuclides strongly exceeds maximum permissible norms. Radioactive strontium has been found in waters of the Pacific Ocean a number of 462 Becquerel's. If radio nuclides get to a food chain, then strontium which is capable to collect in a human body, can increase the risk of diseases of leukemia and cancer of bones.

In Japan, traces of radiation have been found in rice, meat, fruit, vegetables, milk and baby food. All this causes attacks of fear and panic in the population and a heavy burden lay down on the Japanese economy. In January 2012, the Ministry of Economics, trade and the industry of Japan (Ministry of Economy, Trade, and Industry) admitted that radioactive gravel was used for the construction of new houses, repair of roads and other infrastructure injured with an earthquake. Any rules for radiation control of building materials (a stone, gravel), it was accepted not.

Houses, schools, municipal lands need deactivation, up to soil replacement. Still, hasn't solved Japan where it will be possible to store radioactive soil.

ATOMIC WEAPONS

Destructive opportunities of atomic weapons surpass all known before his emergence. It also is a powerful source of radiation and creation of ultrahigh temperatures.

In the military purposes, it is possible to use bombs, shells, torpedoes.

Atomic bomb

Soon after successful, splitting of atom and discovery of atomic energy, in a number of the developed countries, there was a question of a possibility of her use in the military purposes. Large scientists-nuclear physicists in Germany, the Soviet Union, and the USA have been connected to the solution of the specified question. Work was carried out in the conditions of the strictest privacy.

During this period of time for the Nazi regime of Germany was logical to try to use these opening for the military.

The most far-sighted scientists-physicists Hungarians Leo Szilard   (1898-1964) and Edward Teller (1908-2003) perfectly understood it. Forced to immigrate to the USA, they have decided to discuss the fears with Albert Einstein. The great physicist represented, and then the creation of atomic weapons threatens mankind. Therefore it is necessary to outstrip and stop fascist Germany.

The decision to address 32 U.S. President Franklin Delano Roosevelt (1982-1945) with the offer to develop in the country of work on a creation of nuclear weapon has been in common made.

On August 2, 1939, A. Einstein has sent F. Roosevelt the letter in which need to begin work on the creation of an atomic bomb has convincingly been presented. According to instructions of the U.S. President, the special division of engineering troops in the desert (Los Alamos, the State of New Mexico) is created on August 13, 1942. The project has received the name Manhattan.

The brigade general of engineering troops L. Grows (1896-1970) is appointed the head, and the research supervisor - Julius Robert Oppenheimer (1904-1967). By July 1945 the atomic bomb of force unprecedented before has been created.

The beginning of the history of the creation of an atomic bomb in the Soviet Union can be considered 1946. Then the outstanding erudite academician Igor Kurchatov (1903-1960) discussed with I. Stalin a situation on the so-called "Uranium project". In August 1949 test of an atomic bomb is carried out.

It is interesting to note that some scientists-nuclear physicists with horror thought to what hardest consequences for mankind there can be their discoveries. One of the creators of an atomic bomb professor Yu.Oppenheimer has said: "All subsequent generations of people, the admiring discovery of thermonuclear energy in the 20th century, obviously will never cease to damn that time when the creation of human mind has been subordinated to the purposes of the creation of the terrible weapon of destruction".

Later G. Winner wrote: "Atomic bombing – fear of mankind of discoveries".

Not less seriously it has been apprehended by some politicians. So, the Prime minister of England U. Churchill (1864 - 1965) has told:" The Stone Age can return on the shining science wings".

Now the USA, Russia, England, France, China, India, Pakistan, North Korea have officially atomic arms. The big danger is constituted by the accumulation of atomic potential in Iran. The list is replenished both according to official data and according to investigate.

Quite recently Ukraine has declared determination of atomic arms. The specified information isn't submitted convincing. In Ukraine, intensive developments on the creation of nuclear weapon which is allegedly wanted to be used against Russia have begun. Develop an atomic bomb of two types – classical and dirty. The second differs in what infects the territory of the opponent with radiation. The political scientist has emphasized that the West has allowed Kiev "to commit any crimes". The Ukrainian authorities this permission can use and send an atomic bomb to the Crimea or Rostov.

Despite the repeated ban, the race of atomic arms continues and accrues, including the new countries (Iran, etc.).

Atomic bombs can be detonated in the air, on the earth and underground. The explosion of a bomb on the earth is most dangerous. Quite recently in the USA, the project of an atomic bomb is the developed for the destruction of bunkers. It is impossible to exclude that in the future possibilities of use of atomic arms will extend (atomic torpedoes, etc.).

What is represented by an atomic bomb which tragic result of use is already known to mankind?

Depending on filler and e. the substance causing radioactive decay. Distinguish three types of atomic bombs:

- o   uranium or plutonium;
- o   helium and hydrogen;
- o   thermonuclear.

As it is stated above, the effect of atomic bombs includes first of all huge destructive force, equal which in the world doesn't exist. The big danger of an atomic bomb is constituted by high temperature (over 2000 - 3000 degrees powered by Celsius, and ultra-violet radiation. The radiation effect is 5%; the residual radiation is equal to 5-10%. As a result of the explosion of a bomb appear alpha and beta - scale - and ultra-violet radiation. An alpha - and a beta rays are adsorbed by air and don't reach the Earth's surface. The listed effects cause specific changes in an organism – injuries, burns, radiation impact. Their combination fraught with heavier damages of victims is possible. Death is quite possible also at an influence of one of the listed factors.

For the first time in the world, the atomic bomb has been used by the USA at the end of World War II.

On August 1, 1945, the Japanese city of Nagasaki which population made 195 thousand has undergone atomic bombing. Five days later the same fate has comprehended the city of Hiroshima with the population of 255 thousand. As a result of the atomic bombing, many buildings are destroyed. The death toll in Nagasaki has made 39 thousand, in Hiroshima – 66 thousand.

Atomic bomb "Fat man" dumped to Nagasaki

Atomic bomb "Little boy" dumped to Hiroshima

Tens of thousands of people have got severe wounds, burns and radiation damages. Among victims, there were citizens of other countries, including the former Soviet Union.

It should be noted that the presented figures differ according to various sources.

Fungous cloud after explosion of an atomic bomb

It is established that the vast majority of death was caused a shock wave.

As it is stated above, the effect of atomic bombs includes first of all huge destructive force, equal which in the world doesn't exist. The big danger of an atomic bomb is constituted by high temperature (over 2000 - 3000 degree Celsius), and ultra-violet radiation. The radiation effect is 5%, the residual radiation is equal to 5-10%. As a result of the explosion of a bomb appear alpha and beta - scale - and neutron rays. An alpha - and beta rays are adsorbed by air and don't reach the Earth's surface. The listed effects cause specific changes in an organism – injuries, burns, radioactive effects. Their combination fraught with heavier damages of victims is possible. Death is quite possible also under the influence of one of the listed factors.

For the first time in the world, the atomic bomb has been used by the USA at the end of World War II.

On August 1, 1945, the Japanese city of Nagasaki which population made 195 thousand has undergone atomic bombing. Five days later the same fate has comprehended the city of Hiroshima with the population of 255 thousand.

Memorial to the dead in atomic bombing

In memory of a terrible tragedy many memorials in which names of the dead are listed

In anniversaries over monuments and memorials paper cranes whose quantity corresponds to a death toll release. On many monuments there is very touching inscription: "You sleep peacefully, the mistake won't repeat". Near the former epicenter of a bomb National parks and the World museum are created.

Fatal cases from the direct in fluencies of high doses of radiation, according to various authors, approximately have made in Nagasaki two-four thousand people, in Hiroshima – five-six thousand.

In a consequence, the carried-out calculations have shown that the survived persons have received grays loading of equal 0.2 Gr.

To the huge happiness, the world after a bombing of the Japanese cities has no data on the impact of atomic bombs on a health of the person.

However, unfortunately, this trouble can't be excluded in the future.

The radioactive cloud formed as a result of atomic explosion reminded the huge flaring mushroom.

In 1990 the authoritative international organizations have made the decision on non-proliferation of atomic arms and the termination of tests. There is a strong wish to believe in the prudence of the governments of the countries having atomic arms and an active position of the people. At the same time, it should be noted a possibility of the so-called casual emergence of atomic warfare. Such tragic accident can be caused by several reasons - errors of system of information processing and fighting management, technical failures and malfunctions in fighting systems, wrong or unauthorized actions and also nervous breakdowns

Tests of hydrogen bombs

It is known that destructive action of a hydrogen bomb many times over exceeds that of an atomic bomb. In 1954 on the Marshall Islands (USA) the 17th megaton hydrogen bomb has been tested. The explosion of a bomb has happened underground and has extended to 100 miles (160 km.). The direct influence of radiation, a blast wave, and the high temperature wasn't. Nevertheless, increase in a natural radiation background is registered. Long-term monitoring of the area is at once begun.

Hydrogen bomb

Due to a possibility of emission of the radioactive materials which are saved up in the earth the population on is always evacuated in a clean zone. The settled-out persons had no next consequences of radioactive effects. In the next years, the only increase in cases of a hypothyroidism and cancer of a thyroid gland at children is revealed. Within 10 years among observed persons, there was no case of leukemia. Observations continue.

N. Khrushchev has accepted to show the decision on creation of a super bomb in the Soviet Union of "ABBA" to imperialists that we are able to do". The sizes of a super bomb impressed.

An explosion has happened on October 30, 1961, at 11:32 Moscow time. The flash was so bright that it was possible to observe it from distance to 1000 km. Eyewitnesses characterized explosion as the brightest flash on 300-kilometer removal. Later they heard a far and powerful roar. Light of flash proceeded from a huge fiery sphere. Despite the considerable height (4 km), having reached the earth, the fiery sphere continued to grow to the sizes about 10 km in diameter. On this place, there was an orange sphere of the heated gases which has absorbed tens of kilometers of space. Huge "mushroom" has risen to the height of 65 kilometers. After the explosion because of ionization of the atmosphere for 40 minutes, the broadcast message with Novaya Zemlya has been interrupted. At the power of 50 Mega-tone, a zone of elimination represented a circle in 25 kilometers. In a 40-kilometer zone, wooden collapsed and stone houses were strongly damaged. Windows, doors, roofs broke also at long distances. In 60 kilometer from light radiation, it was possible to get burns of the third degree with necrosis of the upper layers of the skin.

Confidential tests of hydrogen and atomic bombs repeatedly took place in other countries including in the Soviet Union. However, this information is a little lit.

Recently North Korea has tested a new hydrogen bomb.

"Dirty bomb"

It is made naturally that terrorists are interested in a possibility of use of radiation for the harmful purposes. This terrible weapon of terrorism.

According to the conclusion of the leading American scientists (T Taylor-1908-2003, Eugene Yester - 1890-1978, and others) use the radioactive weapon in the terrorist purposes is quite real. However, not all want to agree with the specified opinion. It is irresponsible and very dangerous. The so-called dirty bomb can be the most real option of radioactive terrorism. According to some information, the probability of radiation terrorism for the next 10 years is 19%, a use of "a dirty bomb" - 40%.

By the method, less reliable for terrorism, dispersion of radioactive materials or beam influence is represented. Calculations have shown that the small nuclear device which is blown up in the afternoon in New York could become a cause of death of 500 thousand people.

What is represented by a "dirty bomb"? It can be considered the option of an atomic bomb. This combination of explosive material (for example, dynamite) and radioactive material. For radioactive effects Cobalt - 192, Strontium - 90, Plutonium-238, Americium-241 and other radioactive elements can be used.

The sizes of "dirty bombs" vary from big to small which can easily be located in a bag, a portfolio, etc. Production of "a dirty bomb" doesn't present special difficulties since her components, including radioactive materials, can fall into hands of terrorists from the considerable quantity of sources (laboratories, nuclear power plants, radioactive waste, etc.).

"Dirty bomb"

"Dirty bombs" can be used for the defeat of small objects (houses, the subway, stadiums, etc.). Presumably their application for pollution of reservoirs, food objects, etc. Possible consequences of the use of "dirty bombs" depend on many factors (quantity and quality of radioactive materials, the surrounding conditions, weather determining by distribution opportunities, etc.).

The important value has distance from the place of the explosion - it more, the action of radiation less damaging. This complex problem is subject to further deep and comprehensive study.

As well as from other sources of radiation, at an explosion of a bomb radioactive substance can get into a human body from the air through airways, with food through the digestive tract and also through the injured skin.

As a result of the explosion of a 'dirty bomb" perhaps internal and external radiation. Developing of sharp radiation sickness is probable. The big danger is constituted by infection of objects of the environment (means of communication, foodstuff, etc.).

In the latest time, the possibility of radioactive infection of food and reservoirs is discussed.

In case of use of "a dirty bomb" long clarification of the polluted territory for the purpose of prevention of the subsequent influence is of great importance. Believe that the subsequent chronic impact of radiation on an organism has more serious consequences, than during explosion.

It is calculated that finding of the person in the next place from explosion increases a possibility of development of cancer in the following years. Big expenses in communication by the need of evacuation of the population from 78 - a kilometer zone and deactivation of the area are noted.

Bewilderment causes a little reasonable opinion of some Russian, Ukrainian and Belarusian scientists that action of "a dirty bomb" comes down only to the creation of panic among the population ("weapon of mass panic"). Unfortunately, such opinion finds support of some government agencies and is irresponsibly duplicated by mass media

The governments of a number of the countries, including the USA, pay much attention and undertake concrete measures of the fight against atomic terrorism. Recently the UN has approved the decision of the Ant radiation convention on the ban of the use of radiation in the terrorist purposes. There is a wish to hope for strict implementation of similar decisions.

There are some data on preparation for use of "dirty bombs" (London, 2002).

In 2005 in the USA the terrorist preparing explosion of "a dirty bomb" has been detained. Information isn't limited to these facts. Terrorists report about the specific use terms of "dirty bombs".

For the purpose of terrorism, can be used also other isotopes. So, in November 2006 polonium has been poisoned with a radioisotope running the abroad being employee KGB Alexander Litvinenko. The dose received by him made 42 Gray. The victim has died in 22 days.

CHELYABINSK ATOMIC ACCIDENT

In 1945 in the former Soviet Union (in 96 km from the city of Chelyabinsk, in the town of Kyshtym, Urals) the atomic Mayik complex is constructed. Receiving the enriched uranium and plutonium for the military purposes was the main task of his construction. Knew about the existence of complex only separate high-ranking officials of government and military institutions.

The city which has grown around within 45 years was closed. It meant that entrance is strictly forbidden persons and foreigner's unemployed there. The people working and living in it were under constant observation of KGB.

Since 1948 do1992 the amount of the received plutonium was years okolo73 tons. From the beginning of operation on the nuclear reactor periodically there were small accidents. However, no official information existed.

On September 26, 1957, on the nuclear reactor, there was a big accident. The imperfection of a design of the reactor and insufficiently high training of personnel was her reason

Even before the accident, radiation waste was dumped into the river sticking also nearby located lake Karachay. As a result of the accident, the bulk of radioactive plutonium has got to the river Sticking also the lake Karachay. The lake has soon turned into the swamp that has increased transition of radiation to the atmosphere. In total, about 108.000 Curies of radioactive materials are allocated. The bulk of radio nuclides were made by long-living strontium – 90. Radioactive materials were in three physical shapes – firm, gaseous and liquid

CHELYABINSK ATOMIC COMPLEX MAYIK

The polluted territory was protected with a barbed wire. Evacuation of the population from the most infected sites has begun only in 7-10 days. To any accurate information on the incident, measures of protection and prevention of radiation damages weren't.

Everything was deeply coded. Even indicators of blood test and marrow at victims were strictly secret till 1992.

As a result of the accident by high doses (of 0:35 μ/Sv up to 1.7 Stars) about 124 thousand people have undergone radiation. Sticking the radioactivity of samples of water from the small river I made 400 thousand Bq for caesium-137, soils - 120 thousand for strontium-90. A significant increase in a radiation background has been noted in close located districts of the Chelyabinsk and Tyumen regions.

Teenagers ("young liquidators") took part in work on recovery from the accident. Neutralization of the territory isn't complete so far. Consider that it is the region which is the most infected with radiation in the world.

For all years a job of the reactor of 10 thousand people was got by high doses of radiation, from them 4 thousand had that or other severity of radiation sickness. Sharp radiation sickness has resulted from the last most major accident at 42 people. Chronic radiation sickness is found in 1380 persons. 200 people had had heavy radiation burns. In the following years, the plutonium pneumosclerosis is diagnosed for 123 irradiated persons.

Out of time published figures of radiation, pollution was difficult to believe. They were represented overestimated until have got acquainted with the documentary about life and work of victims (2005).

Similar information began to be declassified only in 1992, i.e. in 35 years after the accident. Before practically nothing has been made for prevention and treatment of radiation injuries. The held events are absolutely insufficient to remain to this day. The environment is polluted and on it people, besides, continue to live in very bad social conditions. The resettlement offered by governmental bodies is absolutely unreal. Observation over victims is carried out far not in full. Issues of medical and social rehabilitation are very badly resolved.

Only in April 2005, local authorities have decided to close an object Mayik. Unfortunately, it doesn't solve a security concern and all difficulties which have collected for many years of a lie and divergence. Everything that became an object Mayik, it is necessary to qualify as the fact of use of atomic energy for the military purposes. Now it became obvious to a big circle, both experts, and the population.

RADIATION (RADIOACTIVE) WASTE

The danger of radioactive waste to the health of the person quickly enough increases around the world including in the USA. It is connected as with the steady growth of the use of radiation in many spheres of life, and great difficulties of their clarification, transportation, and burial.

Radioactive waste, as well as other sources of radiation, happens in three physical shapes – firm, liquid and gaseous. The majority of Programs for the elimination of radioactive waste is bad or isn't carried out at all. The sad superiority in this problem is occupied by Russia. It is called "dump" of radioactive waste.

Sources of radioactive waste with various radiation level, in that or another degree, are all branches of science and technicians in whom isotopes, war industry, nuclear power plants are applied. In the structure of radioactive waste tests of atomic weapons and nuclear submarines figure prominently. Highly radioactive waste is formed at nuclear fuel reprocessing and of the fulfilled fuel of nuclear reactors.

Radiation of waste can get into a human body in several ways. Treat them inhalation of the weighed aerosol particles and vapors of water; external radiation, through the digestive tract with food.

Container for transportation radioactive waste

As it is noted, great difficulties are presented by neutralization, transportation and reliable protection of radioactive waste. Various methods of their processing (hardening, verification, cementation, etc.). Facilitating transportation is used.

Recently in the USA for neutralization of radioactive waste have begun to use microbes under the name кineокоkus. As a lot of things in science, properties of these microbes have been found accidentally. Their wide use is very perspective. Nevertheless, the problem of neutralization of radioactive waste is far from the decision.

Not smaller difficulties are presented by transportation of radioactive waste by different types of transport (sea, air, and railway, automobile).

Storage and transportation of radioactive waste are carried out in special containers or in the places excluding their withdrawal an exit of radioactive materials to the environment. It is carried out in 3-5 years.

The huge danger of radioactive waste is defined as an adverse effect on the person and the surrounding nature as it is stated above, a possibility of their use in the terrorist purposes.

In the USA under systematic control, there are zones of radioactive waste which have appeared on places of application of radiation. Control includes regular ecological actions (measurements of levels, cleaning of places of pollution, etc.).

Very great value is attached to studying of the state of health of the population and improving actions in zones of the finding of radioactive waste. In the State of Texas, to the north of the city of Amarillo, the zone in which in the 50th years of the left century passed tests of atomic weapons is located. All next years the region is controlled by various experts, including radio physicians. The central place in this work is taken by studying of the state of health on many indicators.

In the State of Ohio in 52 kilometers from the city, the Columbus is a zone with the increased radiation level in the place of nuclear power plant. Along with ecological actions, numerous Programs for studying the health of the population work.

Scientists of some radioactive materials were given in the State of Colorado in 17 kilometers to the northwest of the city of Denver. By 1992 the territory has been cleaned. All this time and now precede medical observation of the persons living in the region.

Casuistic cases in which ridiculous circumstances were the reason for the radioactive defeat of people are known. So, for example, collectors of scrap metal have got into the thrown clinic of the Brazilian city and have stolen a highly radioactive detail of the equipment used for the treatment of the oncological patients. It has appeared 20-gram the capsule with radioactive cesium. By the number of participants contents of the capsule were distributed in four parts. As a result, the high dose of radiation was received by 20 people from whom four have died. 229 more persons have received easy radiation. In the course of deactivation, it was necessary to demolish 85 houses and to pay huge compensation to the victim.

Similar cases took place and in other regions. Therefore reliable protection of objects of storage of radioactive waste is represented very importantly.

Curies today the problem of transportation and storage of radiation waste is far from the decision.

Recently the International group of scientists has offered a technique of the choice of optimum technologies of processing and burial of radioactive waste. The choice of optimal technical solution is specific and corresponds to the needs for address each case with waste. The offered approach can be extended to treatment of radioactive waste (Russian joint stock Company) from use of nuclear materials, scientific research, power, a nuclear fuel cycle and a conclusion from the operation of nuclear objects and also with the waste containing natural radio nuclides.

Before the choice of the concrete technology of the address with Russian Joint Stock Company, the analysis of the formation of waste, their properties, types, and volumes is necessary. Besides, it is fully necessary to observe regulatory requirements and to provide the existence of the decision on burial provided that his standard legal support exists or it will be established. The choice of technology has to be based on the assessment of all corresponding criteria and restrictions. Detailed information on this subject will be provided in the preparing publication of IAEA with the working name "The Choice of specific technology according to the treatment of Radioactive Waste".

IAEA has developed the international and recognized system of classification of waste which defines the following classes of Russian joint stock Company depending on activity and a half-life period of radio nuclides. Results of researches are studied and complemented.

MECHANISMS OF THE DAMAGING ACTION OF RADIATION

Scientists of the damaging action of radiation on live organisms in experimental conditions have more than centenary history. As a result of this lot of work, some of his mechanisms are defined.

A bit later studying the impact of radiation on the person is begun. At the beginning of observation were carried out over the affected persons working at the objects using radioactive materials.

In the next years, realities of the radioactive defeat of the person increased and will increase. It is connected with the broad use of radioactive materials in many branches of science and practicing, the creation of new technologies, etc.

It is obvious that the specified circumstance has defined need of multilateral and intensive studying of the specified relevant and extremely complex problem. The available data allow gaining certain notion of mechanisms of radioactive effects on the person.

Lately, possibilities of similar scientists steadily increased that became real in connection with progress in many fields of science and technology (physics, chemistry, normal and pathological). Long ago it is noted that the damaging effect of radiation in many respects depends on three factors - duration of influence, distance from a source and its protection. physiology and many others). The damage rate of external radiation is directly proportional to its duration, is inversely proportional to the distance from a source and thickness of a protective layer. In other words, it is more than a distance from a source of radiation and time of his influence, the effect less damaging is shorter. And, at last, the protection of a source of radiation is more adequate and more reliable; the fewer isotopes get to the environment and, therefore, the radiation dose decreases.

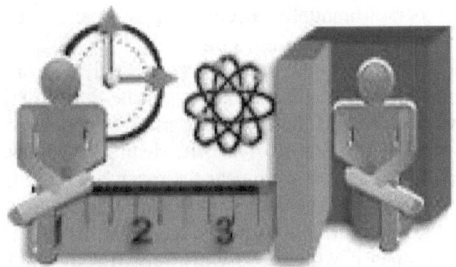

The factors defining the damaging influence of radiation

Reduction of the time spent for a source of radiation is of especially great importance at existence in him the X-ray and gamma ray. It is established that increase in distance from radiation source twice, reduces radioactive effects four times. It gains the special importance at the influence of gamma beams which extend to bigger distance, than an alpha - and beta beams. The protection of a source of radiation, the effect less damaging is more perfect, especially at a beta - and gamma rays.

In the first years after the atomic bombing of the Japanese cities, the distance from explosion epicenter was the main criterion of the extent of the damaging action of radiation. Was considering that the distance is more, the human body damage rate is less. A similar assessment was sometimes used in cases of some accidents. At the same time, the unevenness of distribution radioactive substance couldn't be considered. Now their distribution on the earth is compared to kangaroo jumps.

The specified estimates of radioactive effects were succeeded by more reliable methods of the quantitative and qualitative characteristic of radiation.

The main characteristic of the influence of radiation on an object, including on a human body, is the absorbed radiation dose. She represents the difference of beam energy on an entrance and an exit from that or other environment. Her direct definition is extremely difficult. Therefore in the conditions of practice, indirect methods of definition of the absorbed dose are used. Counters of particles of radiation, ionization chambers, and various types of dosimeters concern them. Possibilities of the retrospective definition of the doses received by the person are intensively studied. The action of radiation on the health of the person in many respects depends on a way of the hit of isotopes to an organism.

Distinguish external and internal radiation. Their combination is possible. External radiation occurs at the stay of the person in the conditions of increase in radiation (air, the soil, reservoirs, etc.). Source of external radioactive infection is not only the surrounding environment. It happens at contact with clothes, footwear, etc. of the irradiated persons.

Internal radiation arises at hit in inside organism of radiation from the   air through airways (an aspiration way), or (and) through a gastrointestinal tract with the infected food and water.

Internal radiation through wound surfaces is very dangerous. Internal radiation, in comparison with external, is always fraught with more serious consequences. It is connected with the fact that isotopes selectively communicate cages. Appropriate authorities also make the long impact. The isotope which has got in an organism is removed not at once and not quickly breaks up, continuing to have the destroying effect in an organism. The danger an alpha - and beta beams are especially high. They have very high density of ionization which is practically absent at impact on the uninjured skin.

Considerable achievement of radiation biology and medicine at the beginning in an experiment, and then at injured people, was establishment of a certain relationship of a dose of radiation exposure and her damaging results

The natural radiation background existing everywhere and also some medical procedures lead to the fact that each person annually receives on average an equivalent dose of radiation from 2 µ/Sv to 5 µ/Sv.

For the people who are professionally connected with radioactive materials, the annual equivalent dose shouldn't exceed 20 µSv· lethal the dose in 8 Sv is considered, and the dose of half survival at which a half of the irradiated group of people perishes makes 4-5Sv.

For 2006 follows from the report of World Health Organization that on the Chernobyl NPP about one thousand people who were near the reactor at the time of accident have received doses from 2 to 20 Sievert that in some cases was deadly.

At liquidators, the average dose of radiation has made about 120 µSv.

High quantity of radiation which is usually received by the patient as a result of a computer tomography of all body is approximately equivalent to the total dose accumulated in 20 years by residents of poorly infected areas after the Chernobyl accident.

At the size of radiation of 10 thousand. (100 Gr) death comes in several hours or days owing to the injury of the central nervous system and others the important of bodies.

The radiation equal 10 000-5000 Rad (100-50 Gr.), leads to death in one-two weeks. Death generally is caused by internal bleedings in the digestive tract.

50% of the irradiated persons perish within 1-2 months at a dose 300-500 is thyroid gland (3-5 Gr). Death is caused by the defeat of all departments of marrow.

Sharp radiation sickness develops at radiation 150-200 Rad (1.5 - 2.0 Gr). Short-term loss of reproduction of posterity occurs at the persons who have received radiation equal 100.

Dose less than 100 Rad (1 Gr) causes a so-called syndrome of bone marrow. It is characterized by damage to marrow, spleen, and lymph nodes. As a result, internal bleeding, weakness, a bacterial infection develops body temperature increases. At influence of the specified dose, the gastric intestinal syndrome can develop. Treat his manifestations: nausea, diarrhea, vomiting, dehydration, violation of electrolytic balance, the bleeding ulcers and also are higher the listed signs of damage of bone marrow.

At a dose less than 5000 Rad (50 Gr.), there is a syndrome of the central nervous system: violation of balance, consciousness, development of spasms also comes to the end he with coma. Usually, at the same time, the specified signs of the defeat of bodies of blood formation, a stomach, and small intestine are observed above. Sharp external influence in a dose from 200 to 300 Gr) causes reddening of integuments, reminding solar burn, and loss

The dose of radiation 125-200 Rad (1.2-2.0 Gr) can cause that or other duration, violation of a menstrual cycle. It is observed at 1/2 irradiated women. A dose in 60 Rad (0.6 Gr) leads to constant infertility.

At a dose 50 l00 Rad (0.5 Gr.) there can be benign tumors of the thyroid gland.

Radiation dose equal 25 Rad (0.25 Gr) becomes dangerous at various burdening circumstances (the general serious condition, diseases of a liver and kidneys, etc.). Doubling of gene mutations, i.e. possibility of congenital diseases is possible at radiation equal 0.01 Gr.

The result of radiation is in many respects caused by various radio sensitivity of bodies. So, for example, temporary violations in genital bodies, marrow, and digestive tract arise at doses at 15-20 times smaller, than changes on the skin. The containing DNA (deoxynucleic acid)). Dose less than 100 Rad (1 Gr) causes a so-called syndrome of marrow. He is characterized by damage to marrow, spleen, and lymph nodes. As a result, internal bleeding, weakness, a bacterial infection develops body temperature increases. At influence of the specified dose, the gastric intestinal syndrome can develop. Treat his manifestations: nausea, diarrhea, vomiting, dehydration, violation of electrolytic balance, the bleeding ulcers and also are higher the listed signs of damage of bone marrow.

At a dose less than 5000 Rad (50 Gr.) there is a syndrome of the central nervous system: violation of balance, consciousness, development of spasms also comes to the end he with coma. Usually, at the same time, the specified signs of the defeat of blood formation, a stomach, and small intestine are observed above. Sharp external influence in a dose from 200 to 300 Gr, causes reddening of integuments, reminding solar burn, and loss acid) is most sensitive to the action of radiation of a cell of fabrics. The submitted data are used for the diagnosis and the prognosis of radioactive effects, holding treatment-and-prophylactic thyroid and the subsequent long-term observation over victims and them descendants more and more widely

Main mechanisms of biological influence of radiation

Negative role incorporated (introduced) in certain bodies radioactive materials (for example, iodine in thyroid gland). As it is stated above, they get to an organism with water food, the inhaled air and through the damaged skin or mucous covers. The combination of the specified ways of the hit of radioactive materials is possible (for example, in respiratory and digestive ways).

To the present time levels of the damaging action of radiation – from a molecule to a complete organism and populations are studied

Negative role incorporated (introduced) in certain bodies radioactive materials (for example, iodine in the thyroid gland). As it is stated above, they get to an organism with water food, the inhaled air and through the damaged skin or mucous covers. The combination of the specified ways of the hit of radioactive materials is possible (for example, in respiratory and digestive ways).

To the present time levels of the damaging action of radiation – from a molecule to a complete organism and populations are studied.

The negative influence of radiation accidents on population distinguishes her from all other accidents (the fire, a flood, an earthquake, etc.) at which population doesn't suffer.

For assessment of radioactive effects on the person quite often.

Level	The main mechanisms of biological effects
Molecules	violation of biological mechanisms, damage, enzymes
Population	change of genetic code
Submolecules	damage to membranes, nuclei, chromosomes
Sells	violation of division, death, transformation, transformation into a tumor
Organs	brain, intestines, bone marrow
Whole organism	death

Table 2. Levels of the damaging action of radiation and its consequence

Use the following two concepts. These are so-called stochastic (Stochastic Health Effects) and not scholastic (No Stochastic Health Effects) effects of radioactive effects.

The stochastic effect is associated with the long influence of radiation of low level. Similar to a situation breaks many processes in an organism which result can be a development of cancer and gene violations.

The not stochastic affect this short-term influence of doses of radiation leading to sharp radiation sickness. It should be noted that radiation damages after single influence appear in the most various terms after radiation. Their manifestations are extremely diverse.

Allocate a so-called radiation syndrome: indisposition, weight reduction of a body, violation of the central nervous system, kidneys, liver and digestive tract, skin changes, inflammation of a warm bag (pericarditis). Besides, quite often there are inflammatory diseases of lungs, sexual dysfunction, a defeat of sight, delay of physical development in children. It is necessary to emphasize that radio sensitivity of children big, than adults. Besides, the radio sensitivity increases at the reception of antibiotics and chemotherapy. The radio sensitivity of embryos and a fruit is especially high.

Total influence of radiation can increase the smoking, alcohol intake, and influence of chemicals. Besides, the existence of individual radio sensitivity and radio resistance (radio resilience), defining consequences of radiation defeat is revealed. This factor represents the relationship between a dose of radioactive effects and intensity of recovery, adaptive and compensatory reactions in an organism. The specified processes are studied at all levels of an organism.

Scientists of cellular mechanisms of the development of radiation syndromes have revealed new data. Distinguish two main mechanisms of the death of cages – interfusing and reproductive death. Between phases develops as a result of activation and release from liposomes of a significant amount of hydrolytic enzymes with the subsequent damage of organelles and titlist cages. Reproductive death happens in the course of a privy or second post beam mitosis or right after them as a result of irreversible violations of structure of chromosomes and defeat of cellular structures radio toxins. At influence of high doses of radiation, the structural violations happen in any bimolecular. At radiation in rather small doses first of all high-polymeric connections are damaged: nucleonic acids, proteins, lipoproteins, polymeric compounds of carbohydrates. Radiation damages structure of proteins therefore enzymatic and their anti-gene activity is broken. According to modern representations, the molecule of protein can lose one or several electrons or to pass into the excited state, at the same time it becomes unstable and easily dissociates with an education of free radicals. Polypeptide chains and also other processes changing the conformational and chemical structure of a pertinacious molecule.

Primary changes in fats consist of education of free radicals that interact with oxygen, are an emergence source the peroxide of connections. The last in turn can react with fats. Thereforehydro peroxides which are very unstable are formed and in the presence of ions of metals easily break up with education reactionary - active radicals.

Primary changes of structure of carbohydrates are observed at the influence of high and come down to a depolarization and oxidation of polysaccharides. It leads to the disintegration of a hydro carbonic chain and formation of acids and formaldehyde.

In the pathogenesis of radiation injuries violations in exchange of nucleonic acids are of great importance. Being the carrier of genetic information in a cage, nucleonic acids are directly involved in processes of biosynthesis of proteins, reproduction of cages and regeneration of fabrics. The earliest reactions to radiation braking of synthesis of DNA in lymphoid tissue, bone marrow and mucous is among a small intestine. The most part of gaps in DNA and RNA, especially single, is exposed to reparation. Double gaps don't repair.

The presented materials testify to many factors of mechanisms and consequences of radioactive effects. Mechanisms of the influence of radiation on a human body are extremely diverse and depend on a number of external and internal factors (Scheme 1).

The biological effect of radiation begins with ionization, i.e. change of a charge of atoms of cages of an organism.

The basic chemical elements of an organism are carbon, oxygen, hydrogen, and sulfur. Oxygen carries out a leading role in splitting of carbohydrates and fats as power sources. This energy is used by cages for the creation of the proteins necessary for the formation of fabrics and also the enzymes which are catalysts (accelerators) of biochemical reactions. Ionization of the oxygen which is in large numbers in and out of cages leads to the destruction of other chemical compounds.

The fats and (or) proteins necessary for the normal activity of an organism can be exposed to the negative impact of radiation. Specific changes, the called mutations result.

It contributes to the development of malignant tumors and their transfer by inheritance.

Radioactive radiation is transferred to body tissues. This transmission of energy leads to damage of kernels of various cages, violation of their activity and finally to death.

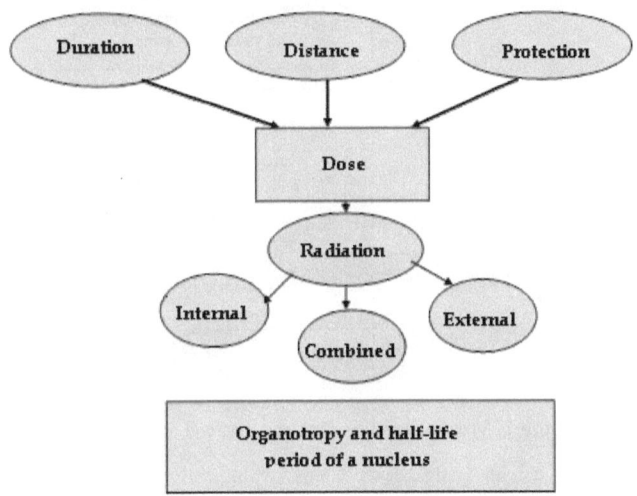

Scheme 1. Characteristic of radiation, defining her damaging action

Dioxynucleinic acid (DNA) (is responsible for a genetic code of each cage. His damage leads to a mutation, i.e. change of the genetic (hereditary) device. The genetic code is a program cage. He is responsible for her division, activity, and function. A marker of genetic disorders is the change of number or structure of chromosomes.

Duration of the listed processes is from one before million seconds.

In the conditions of normal life thin and responsible activities for preservation of kernels of cages are carried out by the special functional system including reparative enzymes. At radioactive effects there comes its insufficiency which, first of all, comes to light in bodies with fast cell fission (marrow, gonads, a small intestine, thyroid gland). The specified processes, in turn, depend not only on a radiation dose. The important value has a condition of the organism irradiated and many other factors.

It is known that the absorbed energy in biological fabrics is distributed unevenly. As a result, the huge amount of energy of radiation is transmitted to certain sites of one cage and absolutely small – to others.

The similar uneven nature of absorption of energy explains a number of features of radioactive effects.

In a human body, there are no barriers to ionizing radiation. Radiation exerts adverse impact on the activity of all biological systems. Besides, oxygen plays a key role in the formation of the substances called by the enzymes which are catalysts for biochemical reactions.

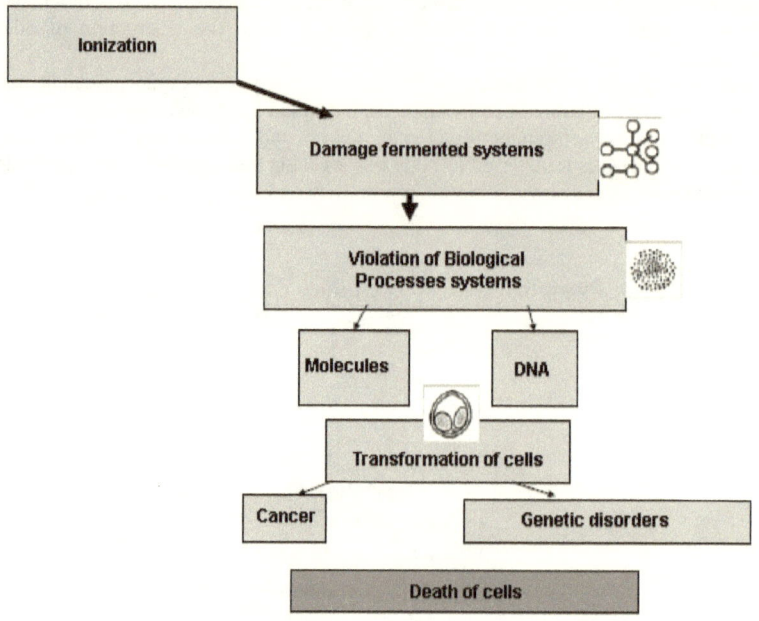

Scheme 2. Mechanisms and consequences of radiation impact on the body

In an experiment and at the irradiated persons the role of so-called free radicals in development of radiation violations profoundly is studied. As a result of radiation damage to molecules the number of free radicals increases. They, in turn, exert the damaging impact on organism cages. Increase in their contents is revealed in the blood and the damaged bodies irradiated that confirm the specified opportunity.

As the reaction to all events, the protective mechanisms directed to correction, adaptation, and compensation turn on. But their opportunities are limited and in process of increase in the absorbed dose, they are exhausted. Then there occur changes in cages of an organism and their death.

In the practical relation the fact that cages of various bodies are unequally sensitive to radioactive effects matters. As it is specified, bodies which cages slopes to fast division and are less specific are most sensitive (for example, bodies of blood formation). Besides, various radioactive materials have a specific influence on cages and body tissues.

To iodine-131 thyroid gland, is most sensitive to ruthenium-103 – the surface of the skin, lungs. Bones and marrow are most sensitive to Strontium-90 and Plutonium-229. Caesium-144 is especially dangerous to lungs and lymphatic system.

The list of bodies and fabrics, highly sensitive to radiation, in process of further researches, will, apparently to increase.

Before the Chernobyl accident mechanisms of the damaging action of radiation at the long influence of small doses have been less studied. A small dose of radiation size to 1 Gr is considered.

A feature of long influence of small doses of radiation is that at the same time more favorable conditions for inclusion so-called the compensatory of mechanisms are created. On the other hand, repeated influences by many kinds of the isotopes are increased by the distribution of defeat by various systems and human organs. The damaging effect of the isotope can remain for years and many decades. Dangerously and the fact that millions of people can be exposed to radiation.

Accident at the Chernobyl nuclear power plant has created conditions for more profound studying of this problem. Within more than 25 years she is studied by prominent scientists and specialized institutions around the world.

It is necessary to emphasize that that or other damaging effects as well as at single radiation, is shown through various periods of time.

Refer a hair loss to the next effects (a dose do10– 0.1 Gr.) to full baldness (a dose to 400 Rad – 4 Gr). The hair reconstruction comes in two months, but their color and quality change. It is necessary to remember that the radioactive effects can be combined with other damaging factors (for example, chemicals). In similar cases, there are additional mechanisms of damage. The data which are available now confirm multidimensionality of radioactive effects on the person.

Studying of mechanisms of the damaging impact of radiation on the person intensively continues. Their further studying will allow improving methods of treatment and prevention of consequences of the influence of radiation.

CONSEQUENCES OF RADIATIVE EFFECTS

In 60-70 years of the left century radio biologists and radio physicians have begun to study not only single direct impact of radiation on an organism but also the remote effects, i.e. diseases mediated by it. They are called the radiation induced (stimulated) diseases. Today more than 20 similar violations and diseases are known. Treat them: cancer, influence on heredity, an immunosuppressant and an immunodeficiency, increase in susceptibility to bacteria and viruses, violations of the endocrine system and a metabolism, developing of a cataract, temporary or constant infertility, reduction of the average expected life expectancy, a delay of mental development and others. Eventually, the list of the radiation stimulated diseases isn't reduced and replenished.

Studying of the listed and other problems are promoted by achievements of many biological and medical sciences.

Above the presented materials demonstrate that the radiation effect at the person is defined by many characteristics (Scheme 1). To it is necessary to add the character of a source of radiation, the way of the hit to an organism and specific features of the persons which were affected. Influence of the previous radiation level of the field of accommodation is also noted. He is higher; the consequences of radioactive effects are less expressed. It is explained by the higher degree of adaptability of an organism.

Radiation damages and diseases I am conditionally divided into two groups: early and late.

Early radiation damages

It is necessary to refer the damages arising in connection with radiation diagnostics, especially to early changes of radioactive effects - radiation therapy and sharp radiation sickness.

The radiation damages connected with diagnostics and treatment

Radiological scientists have gained distribution in all sections of scientific and applied medicine. It should be noted improvement of diagnostic and therapeutic installations, drama for the last decade, methodologies and techniques of their use. Very important calculations of a dose of radiation at the application of various techniques of radiation diagnostics are represented It allows doctors rationally, without harm for the patient, to apply that or other diagnostic method.

Today for these methods there are practically no age restrictions. They are contraindicated in early terms of pregnancy and also at hyper sensibility to separate isotopes (for example, to iodine). The doctor can always obtain the specified information from the patient. Categorically it isn't recommended to the patient to insist on carrying out necessary, from his point of view, additional radiation procedures. It is fraught with a transformation of a safe research into dangerous.

To rare complications of radiation diagnostics, except allergic reactions, belong burns of skin and mucous membranes.

It is known that radiation methods of treatment of various forms of cancer are the most effective and therefore are widely used. It concerns persons of both sexes in all age periods. However, their use can sometimes cause a number of complications.

Ray loading change in the big ranges. As a rule, at patients, those or other signs of sharp develop, is rarer than chronic radiation sickness. Treat them: suppression of function of marrow, beam reactions of a number of bodies (lungs, heart, and digestive tract).

Bleeding, anemia can result from the oppression of functions of bone marrow and also develop infectious and fungal diseases of various bodies. Usually, such patients have changed on the skin (hypostasis, reddening). The hair loss, dysfunction of genitals is possible. Under the influence of special treatment the listed signs usually disappear, but then sometimes appear repeatedly. Recovery can last up to two years.

Use of very high doses of radiation is occasionally fraught also with serious complications. Radiation injuries of a brain, lungs, heart, persistent nausea and vomiting concern them.

Damage to a brain (sharp encephalopathy) is shown by nausea, a headache, and spasms. Her emergence is possible as in the initiation of treatment, so in weeks and months. More often the specified complication arises at beam cancer therapy of a brain. Encephalopathy and radiation damage of bone marrow. It can repeatedly develop in months and even years after the end of treatment.

Impact on the heart is characterized by varying severity of changes of a muscle and a pericardium. At heavy violations the forecast adverse. The frequency of the listed complications increases at patients to whom radiation treatment in combination with chemotherapy is carried out. Treat infrequent complications radiation nephrite, a nephropathy, prostitutes, cystitis, proctitis.

All arising complications immediately are subject to complex treatment. At timely use of modern methods of prevention and treatment at over the whelming number of patients an outcome favorable. The improvement of a technique of treatment allowing reducing beam load of healthy fabrics is important.

Acute Radiation Sickness (ARS)

In the American literature use the name " **Acute Radiation Sickness** more often - ARS (Acute radiation syndrome - ARS), but not sharp radiation sickness (Acute Radiation Disease).

As it is already noted, results of radiation in many respects are defined by his dose and ways of the hit to an organism.

ARS arises at a combination of certain conditions: a high dose, her external influence by the getting beams (a beta - and scale) and short time. At higher doses of radiation instantly there comes death. The high doses of radiation causing ARS are possible at major accidents (a cyclotron, nuclear power plants, radioactive waste), atomic bombings and, perhaps, at the use of "dirty bombs".

At ARS the immune system is surprised, as result resilience to infectious diseases sharply decreases. 50% of victims in the first two months die of infectious diseases.

To 40-x years of the left century cases of ARS were estimated in tens and have been usually connected with labor activity in scientific laboratories. The wide use of radiation energy has defined significant increase (hundreds of cases) in ARS. The dependence and time of emergence of various signs of intense radiation injury on radiation dose are presented in table **3**.

Dose (Gr)	Signs emergence	Time
0.5-1.0	blood changes	hours
0.5 0	nausea	hours
0.55	weakness	hours
0.70	vomiting	hours
075	loss of hair	2-3 weeks
0.90	diarrhea	2-3 weeks
1. 00th	bleedings	2-3 weeks
4. 00th	death	2 months
10. 0	Violations gastrointestinal tract, internal bleedings	1-2 weeks
20.0	defeat central, nervous system, death	minutes, hours, days

Table 3. Time of Emergence of the Main Signs of ARS depending on a radiation dose

The submitted data show that weight of signs and death of ARS are directly proportional to a radiation dose.

All researchers observing over victims after the accident on the CNPP have come to the same conclusion.

After atomic bombing of the Japanese cities and accidents on nuclear power plants, and the, patients were observed in unequal conditions. Therefore the description of ARS and systematization of her signs, in a certain measure, differ. Nevertheless, the main signs in many respects coincide and correspond to results of pilot studies.

As a result of the accident on the CNPP of 140 people with sharp radiation sickness of varying severity were treated in specialized offices of Moscow and Kiev. The radiation dose at them made 350-500 µ/Sv. Depending on weight and an outcome allocate four forms (degrees) of ARS: super sharp, very heavy, heavy and easy.

At a super sharp form quickly there comes death. It was observed after the atomic bombing of the Japanese cities and at accidents on the NPP.

31 employees of the CNPP had the very severe form of ARS. All of them have died in the next weeks after the accident.

At a severe form diarrhea, vomiting, temperature the increase of a body, multiple hemorrhages on the skin and mucous membranes, in internals, intoxication, the decrease in the quantity of leukocytes, including their versions – lymphocytes and also platelets are observed.

Among the population, sharp radiation sickness isn't registered. For the characteristic of the ARS easy form allocation of stages is expedient.

1 stage, so-called prodrom, is characterized by lack of appetite, nausea, vomiting, diarrhea, salivation, colic belly-aches, dehydration is possible. The listed symptoms
arise in a few minutes or hour after radiation and proceed within two days.

2 stage – asymptomatic, i.e. disappearance are higher than the listed signs. It lasts from several days to one week.

3 stage – renewal of the listed symptoms and emergence of signs of severe damage to bone marrow (bleeding, anemia, etc.).

4 stage – the recovery lasting from several weeks to one month.

After the Chernobyl accident during 1987-1998 death has come at 10 people who have received a dose of radiation of 1.3-5.2 Gr. From them three persons have died from a coronary not enough, two – from the severe damage of marrow, two - from cirrhosis, one patient had a gangrene of lungs and tuberculosis. One patient has died of sharp myeloid leukemia.

Heavy radiation burns were observed in 56 patients affected by the Chernobyl accident. Usually, such patients in the subsequent had a cataract

Burns settled down on a face, a neck, brushes, legs, buttocks more often. They appeared earlier and more hard proceeded on the damaged sites of skin. On the uninjured skin, burns arise only at high doses of radiation and direct contact (for example, from clothes).

The famous radiologist *Henri Becquerel* had had a skin burn after in a side pocket of a jacket there was a test tube with radioactive material. Unlike thermal, chemical and other causes, radiation burns develop not at once.

In recent years radiation burns are designated by the term "Skin Radiation Syndrome". He is characterized reddening, dryness, formation of superficial defects. The combination of radiation and thermal burns is possible.

Radiation burns are followed by the cruel pains which aren't conceding to modern methods of treatment. The specified signs sometimes decrease, and then again appear.

In the subsequent on the place of a burn skin coloring changes, there can be consolidations and ulcerations, the growth of keloid fabric.

For determination of severity and the forecast of a skin radiation syndrome, the choice of adequate treatment the thermography, ultratsonnografiya, a biopsy is used.

Quite often there are infectious and fungal inflammations in the area, the injured skin. The skin radiation syndrome is more often observed at heavy degrees of ARS and considerably worsens the forecast.

Less hard cases of radiation burns of skin are characterized by passing reddening and an itch. The listed signs disappear in several weeks. In the subsequent on the damaged sites pigmentation (brown coloring of skin) appears.

Allocate five stages of radiation injuries of skin.
 1. Prodrom arises in 24-72 hours.
 2. Demonstrated, an emergence of symptoms through (days, four weeks).
 3. Sub sharp develops in four-six weeks.
 4. Chronic arises through three one or two months of the year.
 5. Late it is shown in decades.

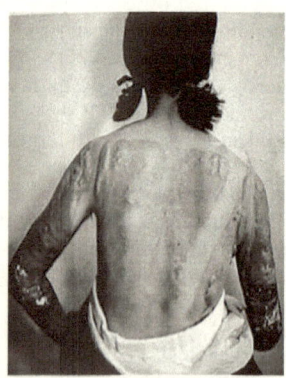

Combination of radiation and thermal burns

At the persons giving first aid irradiated in a zone of the accident of the CNPP (doctors, nurses, etc.), so-called contact burns of the skin (reddening of the skin, an itch, peeling) were observed.

Almost at all patients with the ORS easy form after the accident on the CNPP functional impotence developed.

From the very first days, ARS is observed sharp oppression of immune system which isn't always shown by visible signs.

After the atomic bombing, the Japanese scientists have for the first time noted that men are more susceptible to radiation then women. At women of ARS arises at high doses of radiation then at men. It is explained not so with distinctions of sex hormones, how the much bigger resistance of women to a stress. The specified data don't coincide with observation over victims after the Chernobyl accident, the female persons who have established big sensitivity to radiation.

Death from an acute ray rays syndrome comes in several hours after radiation, at smaller doses – within a month.

Diagnostics of a acute rays syndrome comes easy if a radiation established the fact. It is very important to have information on the received dose since it substantially defines a technique of treatment and the prognosis.

At suspicion on ARS, it is necessary to carry out monitoring, i. e. regular blood test (maintenance of erythrocytes, platelets, leukocyte formula). At opportunity, it is necessary to make the cytogenetic blood test for the definition of chromosomal aberrations. If there are no instructions on radiation exposure, it is necessary to exclude other diseases having similar signs (heavy sepsis, etc.).

Along with a dose signs of defeat are defined also by the   primary accumulation of isotopes in that or other body (intestines, a brain, etc.). It, in turn, depends on the chemical composition and properties of radioisotopes.

As it was specified, the organism of the child is more sensitive to the negative impact of radiation in comparison with adults. Therefore at children, the listed signs of ARS appear at lower doses of radiation, than at adults.

After atomic bombing of the Japanese cities, bleeding of skin, mucous membranes, and hemorrhage in internals were one of the main symptoms of ARS. They appeared as a result of the defeat of certain sites of marrow. In this regard, there came oppression of factors of fibrillation and, first of all, reduction of the formation of the specific cages called platelets. Violation of fibrillation as a result of which there is bleeding was the main reason for the death of victims.

ARS at the persons which have undergone atomic bombing in the Japanese cities has certain features. It is connected by radiation combination to a huge blast wave and a high temperature of the environment. Therefore some researchers find it difficult to speak only about a possibility of the isolated radioactive effects.

The provided information demonstrates that ARS has the certain features caused by a source of radioactive effects (atomic weapons, accidents, etc.) and the received radiation dose.

Death from an acute ray syndrome comes in several hours after radiation, at smaller doses – within a month. Diagnostics of a sharp ray's syndrome comes easy if a radiation established the fact. It is very important to have information on the received dose since it substantially defines a technique of treatment and the prognosis.

At suspicion on ARS, it is necessary to carry out monitoring, t. e regular blood test (maintenance of erythrocytes, platelets, leukocyte formula). At an opportunity, it is necessary to make the cytogenetic blood test for the definition of chromosomal aberrations. If there are no instructions on radiation exposure, it is necessary to exclude other diseases having similar signs (heavy sepsis, etc.).

Late radiation damage

General information
The concept" late radiation damages" isn't defined in concrete terms. It can be months, years, and decades after radiation. Their manifestations are caused by many reasons. The dose, duration and the nature of radiation, isotope half-life period duration, body and the system of defeat, the previous condition of an organism, time and quality of holding treatment-and-prophylactic actions social   indicators etc. concern them. Therefore their systematization presents certain difficulties. It is possible to carry cancer of various localization, congenital and genetic disorders to the most frequent manifestations of late radioactive effects. Late manifestations of radiation can be connected as with her direct influence (cancer, hereditary diseases), and is radiation - the caused defeats of systems of an organism, separate bodies, and processes.

Radiation and cancer

At a dawn of use of radiation connection of emergence of various forms of cancer with radiation has been noted. In the subsequent skin cancer cases at the employees having contact with radiation observe (doctors and laboratory assistants of x-ray offices and others). Leukemia was a cause of death of one of the authors of receiving radioactive materials of Maria Sklodowska - Curie.

Numerous long-term experiments on animals have convincingly confirmed the specified communication. In process of improvement of radiation protection, similar observations have kind of faded into the background.

To replace the long-term epidemiological scientific which has established the connection of radiation with developing of cancer modern methods of molecular biology; biochemistry and other sciences have come. Use of such methods has allowed proving scientifically radiation role in the development of cancer.

It is made naturally that in places of radiation pollution there can be also other factors promoting developing of malignant tumors (for example, some chemicals). In this regard are developed and the indicators demonstrating radiation origin of cancer are already offered. Today establishment of a dose of radiation and some microscopic characteristics of a cancer tumor and also feature of a course of a disease concern them.

Increase in frequency of leukemia and malignant tumors at the children born from mothers passing a radio graphic research is revealed. According to 10 years' observations over 15 million single children and 350 thousand twins who have undergone pre-natal radiation, the risk of a disease of leukemia and cancer increases. Increase in the specified risk at one child increases by 1.5 times, at twins – in 2.2 and 1.6 times.

Radiation accidents of the left century have defined need of deep and comprehensive study of this problem.

After atomic bombing of the Japanese cities increases in the prevalence of various forms and localizations of cancer is noted. The Japanese authors consider that cancer belongs to the most serious consequences of radiation. The frequency of his emergence reveals distinctions in relation to that or another body.

Cancer of thyroid and chest gland is carried to the most common forms of cancer. These forms are diagnosed for 10 persons on 1000 irradiated, i.e. in 1% of cases. Cancer of other bodies was observed at one person on 10 thousand irradiated i.e. in 0.1% of cases.

It is revealed that after the atomic bombing all next year's cancer was the main reason for the death of the population of Japan. Its specific weight in the structure of causes of death increases in process of increase of terms after radiation.

In years after atomic bombing methods of medical statistics have in many respects changed. It, certainly, has exerted the impact on incidence figures, in particular, cancer. However, the determined consistent patterns generally remain. At the end of the left century, huge work in which long-term researches of consequences of influence of radiation on health are summarized and analyzed is carried out. Reliable the increase in prevalence is established for the following forms of cancer: leukemia, except for lymphoid and T-cellular; breast cancer, thyroid gland, large intestine, stomach, lungs, ovaries.

The prevalence of cancer of gullet, salivary glands, a liver, skin, a bladder, nervous system, a myeloma. and a malignant lymphoma has less authentically increased.

Increase in prevalence of chronic lymphoid leukemia, pancreatic cancer, gallbladder, rectum, uterus, bones is subject to further studying. The importance of the increase in the prevalence of the listed forms of cancer can be established at the bigger number of observations.

Curves of cancer cases of various localization by years are various. Her "peaks" differ with time which has gone after radiation. It is emphasized that the persons who have received radiation in the childhood and young age get sick more often.

After the accident on the CNPP bladder cancer cases from 1986 to 2000 have increased in Ukraine with 26.2 to 43.3 by 100 thousand populations. It is revealed what with urine is allocated to 80% of the caesium-137 which has got to an organism. The obtained data explain the growth of incidence of this localization of cancer. On the basis of in-depth studies, in common carried out Ukrainian and Japanese scientists, is made a hypothesis of a role of free radicals in carcinogenesis of a bladder.

In recent years the average annual gain of cancer of various localizations makes among the irradiated persons 5.5%, and in control group of unpredicted – 1.5%. The highest rate of an average annual gain of cancer is observed at the liquidators and persons living in a 30-kilometer zone - 7.9%.

Others the colon cancer, kidneys, a bladder meets more often. Developing of cancer at the younger age, both at men, and at women is noted. Many scientists predict as further radiation will figure prominently among cancer reasons. It should be noted the ambiguous character of the provided data. In this regard observations over victims and their posterity continue and have to continue still very long.

The formulas allowing predicting cancer at the irradiated persons are offered. Their reliability is carefully studied.

Accidents at the atomic enterprise Mayik have for many years caused the growth of incidence of a breast cancer, lungs, and blood, a brain, genitals at men and women. It should be noted that the specific weight of defeat of that, or other body is various.

Terms cancer various localizations developing in different radiation conditions (atomic bombing, nuclear power plants, etc.) differ in a variety.

Communication of some forms of cancer the irradiated persons is noted with age. So, cancer of chest gland mainly arose at female aged up to 40-45 years, and thyroid gland – at children.

It is necessary to emphasize that developing of radiation cancer is possible in many years and predisposition to him it will be descended. Predict that at liquidators of consequences of accidents on nuclear power plants, radiation cancer will make 10% of all cases.

Rather a new section of radio medicine which has appeared after the accident on the CNPP is diseases of the thyroid gland, in particular, cancer.

In the first days and months after the accident in a radio nuclide range considerably prevailed Iodine -131 which place of accumulation is thyroid gland.

Therefore it was the first target of the Chernobyl accident.

Thyroid gland - unpaired body weighing 20-30 gr. located on the forward surface of a neck in a thyroid cartilage. The healthy thyroid gland isn't visible approximately but can be probed on the forward surface of a neck at persons with badly developed hypodermic fatty cellulose. The main function of thyroid gland - development so-called hormone Thyreoidinum participating in many processes of normal life support of an organism.

Cancer of the thyroid gland before the Chernobyl accident in the former Soviet Union met extremely seldom, and at children was practically not observed. These data coincide with publications of the countries of Europe and America. Due to the rare occurrence of cancer of thyroid gland, his diagnostics, prevention, and treatment were a little studied.

The scientist conducted after the atomic bombing have shown that cancer of thyroid gland arises in 6-7 years after radiation and this effect remains 20-30 years. Inspection of the Japanese cities which have survived after the atomic bombing, have shown a linear increase of the frequency of cancer of thyroid gland depending on age and a dose, of radiation. However, at very high doses inverse relation came to light.

Later terms of the emergence of cancer of thyroid gland were observed in adults (11 years later) after test of a hydrogen bomb on the Marshall Islands.

After the Chernobyl accident growth a number of patients with cancer of a thyroid gland have begun in 3-4 years, i.e. in earlier terms, than after other radiation accidents (atomic bombing, a test of a hydrogen bomb, etc.).

The prevalence of cancer of thyroid gland at children in various regions of Belarus has increased at 50-100-200 times, in Ukraine-in 5-7-10 times.

In Russia, not so expressed growth of prevalence of cancer of thyroid gland, took place only in several districts of the polluted areas. Among patients children aged from 5 up to 14 years and irradiated aged from the birth up to four years prevailed. Communication of development of cancer of thyroid gland of radiation is found with age. His frequency decreases in process of increase in age of children. In all age periods, female persons prevail.

Higher incidence at children is connected as with big sensitivity of a thyroid gland to radioactive iodine, and with consumption of dairy products, become in conditions after the accident, the main source of radioactive iodine.

At the same dose of radiation, children get sick considerably more often than adults. It is connected with the fact that thyroid gland at children more accumulates radioactive iodine and as it is stated above, is more sensitive to his influence.

The growth of cancer cases of the thyroid gland which has begun in 1990 at adults and children continued till 1998. Then at the children, it was stabilized, but at the higher level, than before the accident. It is supposed that the following rise in incidence can come in 10 and more years. The risk of a disease will remain for 50 years after the accident. The cancer cases of the thyroid gland are "spasmodic".

Among adults, especially liquidators of consequences of the accident, increase in cancer cases of the thyroid gland has begun later and so far tends to increase. Essential distinctions of cancer cases of the thyroid gland at the settled-out persons and continuing to live in the polluted territories are revealed. Cancer of the thyroid gland at the settled-out persons was observed almost twice less than the polluted territories living on. Most often cancer of thyroid gland came to light at liquidators of the accident.

Observations in regions of radiation accidents in the USA taking place in 40-60 years haven't established the accurate dependence of prevalence of cancer of thyroid gland with radiation. Believe that it is connected with various ways of the hit of radioactive iodine to an organism.

At injured with the Chernobyl accident Iodine - 131 got to an organism in three ways - through the injured skin, respiratory organs and digestion.

It is possible to assume that at victims in the USA has been excluded, or the possibility of a hit of isotopes in respiratory and digestive ways is limited. It is supposed that radiation cancer of a thyroid gland in the in subsequent years will make up to 10% of all cases of diseases at children. The microscopic analysis has established that at the diseased the severe (proliferate) form of cancer prevails. After the long-term discussions, it is proved that radioactive Iodine is the reason of cancer of thyroid gland at observed persons.

What symptoms of cancer of thyroid gland?

In an initial stage of a disease, any manifestations usually are absent. Rather seldom there are signs of a hypothyroidism. Due to the superficial arrangement of the body, its increase can be noticed at survey or palpation. Adults can feel thyroid gland independently (self-inspection). Children need assistance (parents, medical staff, etc.).

It is noted that radiation cancer of thyroid gland differs in rapid growth and metastasis, i.e. distribution to other bodies (lungs, lymph nodes, a liver, etc.). Therefore quite often the diagnosis is established on the basis of signs of spread of cancer, i.e. in far come disease stage. The blood spitting, increase in supraclavicular and axillary's lymph nodes, bone pains, the temperature increase of a body concern them.

Biopsy of thyroid gland – a research under a microscope of a piece or cages of the body. For receiving cages special syringes sucking away, aspiration) a biopsy is used. Results at a research of a piece of tissue of the thyroid gland received in the surgical way (a surgical biopsy) are more reliable.

The listed diagnostics methods in that or other combination are used at observation over persons, being, ever and the radioactive iodine living today in places of pollution.

From this group, it is necessary to allocate the persons which are subject to prime inspection. Treat them:

- persons which have undergone beam influence aged up to 14 years;

- liquidators of consequences of accident;
- peoples who have transferred radiation thyroid ;

The huge value for determination of the sequence of inspection is represented by data of dissymmetry of a thyroid gland. It is revealed that the risk of a disease at the children who have received a high dose of radiation increases six times in comparison with the children who have received the minimum dose. Unfortunately, only a few persons have such data. This gap, to some extent, is compensated by use of calculation and other methods of definition of a dose of the received radiation.

Radiation as cause of infringement of a reproduction, congenital, genetic and other diseases

Violations of growth and development of a fruit under the influence of radiation in the conditions of a scientific experiment are found long ago.

These scientists at people belong to the rather new division of science and practicing called "Radiation genetics." On 1927 is considered the date of the birth. Then the German scientist Herman Miller has for the first time published convincing experimental results about radiation role in genetic disorders.

Observations over people became possible and necessary in connection with the use of an atomic bomb and major accidents which have caused radiation of a large number of people.

Early studies in this area have begun after the atomic bombing of the Japanese cities and proceed so far.

The Chernobyl accident, as well as other accidents, was the basis for the deep and comprehensive study of this problem.

Studying of radioactive effects includes the following aspects:
- birth rate;
- course of pregnancy, childbirth and postnatal period;
- indicators of health of children;
- condition of genital bodies of mother;
- congenital and hereditary diseases at radiation in the embryonic period;
- influence of radiation of one, or both parents, on posterity.

After the Chernobyl accident, there was a sharp decrease in birth rate. Its level in some regions was lower than during World War II. It is in many respects connected with the increase in the quantity of abortions in early terms of pregnancy and abortions.

In the subsequent, it is established that the feta placenta insufficiency resulting from radioactive effects is the reason of abortions.

There are messages about the accumulation of some isotopes in a placenta. Besides, in the first years after the accident, the number of abortions has increased. It has been connected with the fear of parents for the health of children, including developing of congenital and hereditary diseases.

In the last 10 years the birth rate in Ukraine has increased, however, remains below accepted standards. The irradiated women had a pregnancy nephropathy, premature birth, and postnatal complications more often (bleedings, an infection in patrimonial ways, etc.).

Increase in cases of endometriosis and cancer of genitals (ovaries, uterus) is revealed. According to the Belarusian authors, the incidence of endometriosis has increased twice and it arises at women of younger age more often. It in turn leads to the reduction of birth rate. After the atomic bombing, the prevalence of myoma of a uterus has significantly increased.

Especially it is necessary to stop on radioactive effects on a fruit. The negative influence of radiation on a fruit is most sharply expressed within 36 hours after conception.

In the first two weeks of pregnancy, the radiation effect heavy and quite often is a fruit cause of death. Death of a fruit can come still before the woman learns about pregnancy. If the fruit doesn't perish, there is a high probability of development of various congenital defects. Their weight is directly proportional to a dose of radiation and is inversely proportional to pregnancy terms. In other words, the danger of radiation of a fruit decreases with increase in terms of pregnancy.

Between the second and 15th week of pregnancy at high doses of radiation there can be various heavy defects of development of the congenital diseases, in particular, a brain. From 16 to 26 weeks of pregnancy, defects of development arise only at radiation by high doses.

After the 26th week of pregnancy sensitivity of a fruit to radiation approximately such as at newborns, congenital defects don't arise anymore. However, high is a probability of developing of a malignant tumor.

The frequency of various violations at a fruit depending on pregnancy terms in which there was a radiation is calculated. Radiation of an embryo of the person within the first two months leads to 100% to defeat, from three to five – to 64%/, from six to ten to 23%.

Generally, the sensitivity of a fruit to radiation in 10-300 times more in comparison with an adult organism.

Fractional, i.e. repeated radiation by small doses, leads to heavier damages than single high doses. It is connected with the fact that influence is the share of various types of germinal cages. Damage to a large number of rudiments of the bodies which are at critical stages of development results.

The listed violations of a reproduction and other changes (congenital and genetic diseases, etc.) are established in the certain European countries in which increase in a radiation background after the accident on the CNPP is noted (Germany, England, etc.). So, according to the English researchers, congenital diseases in the country have increased by 50%.

According to observations of the Israeli scientists over repatriates from the infected regions of the former Soviet Union, the number of congenital and hereditary diseases has increased seven times.

Almost at a half (45%) of the children who were born from mothers who have undergone radiation in terms of pregnancy of 7-15 weeks signs of intellectual backwardness were observed. Besides, at posterity of the women who have transferred radiation in the first half of pregnancy the reduction of the sizes of a brain, brain dropsy, a growth inhibition, mongolism, congenital heart diseases, uric ways and other bodies is observed.

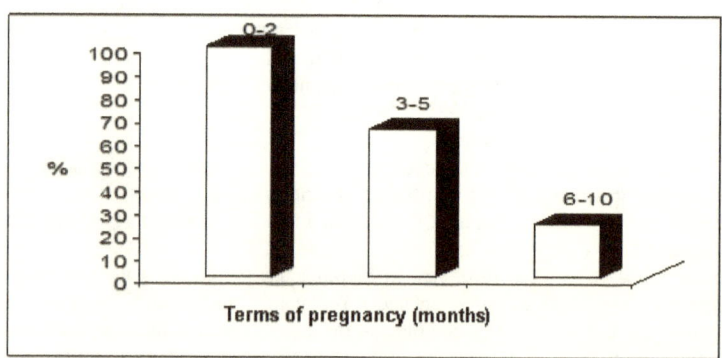

Congenital diseases at radiation in various terms of pregnancy.

The amounts of the congenital defects arising in connection with radiation are several tens of names. Besides listed, changes of a shape of a skull, a funneled breast, dislocations, diseases of teeth, squint, a congenital cataract, diseases of a mesh cover of an eye, glaucoma, damage of teeth, heart diseases, diseases of kidneys, genitals and others are observed

LONG IMPACT OF SMALL DOSES OF RADIATION ON HEALTH

On the basis of the long-term generalized data obtained after atomic bombing the reduction of the sizes of a brain, intellectual backwardness, a growth inhibition and development, bad progress at school, low IQ (intelligence quotient) significantly increase.

Communication of incidence of leukemia and terms of pre-natal and subsequent radiation is calculated.

According to some scientists radiation of an embryo in small doses can cause violations which can't be revealed by modern methods, but which can contribute to the development of changes in many years. The arising violations at the cellular level create a real basis for descended, i.e. genetic diseases.

The presented materials convincingly demonstrate the negative impact of radiation on a reproduction, increase in the prevalence of congenital and genetic diseases. The specified changes can appear both as a result of radiation in the period of an embryonic development, and at radiation of one or both parents. As it has been specified, many changes pass from father to son, i.e. negatively influence the population.

Results of multidimensional and long-term scientific research and also practical observations have allowed improving diagnostics, treatment, and prevention of various early and late radiation damages. Scientists will be continue as early as many years.

In process of expansion and deepening of scientist about the influence of radiation on the person, many scientists have paid attention that it isn't limited to the growth of prevalence of cancer, congenital and genetic diseases.

Accidents on the NPP and other atomic objects, precedence ness of the Chernobyl accident with all evidence have dictated need of studying of different aspects of radioactive effects on the of the person.

Literally from first weeks after the accident on the CNPP comprehensive programs of scientists, taking into account early the obtained data are developed (after the atomic bombing, accidents on the NPP, etc.). They included studying of a condition practically of all systems of an organism. Eventually, the specified scientists extend and go deep.

Medico-demographic indicators

The violations revealed so far and diseases concern the systems of an organism and separate physiological processes (see below). As a result, medico-demographic parameters which include indicators of the health of the population have changed and change. Carry to the standard indicators of the health of the population: life expectancy, incidence and prevalence of diseases, level of birth rate and mortality and also primary disability.

Multidimensional scientists have established that in the regions polluted by radiation life expectancy of the population, especially men have decreased. The prevalence of many diseases has increased. The heavy course for the radiation caused diseases and difficulty of their treatment were the reason of the increase in mortality and reduction of life expectancy. According to the Ukrainian sources of literature in 10 years after the accident of life expectancy of men was lower than in 20 poorest countries of the world. During the same period at 84% of persons, having undergone radioactive effects, and at 92% of liquidators those or other diseases were observed.

Changes in the specified indicators in many respects were defined by the extent of radioactive pollution. So, in the Gomel region (Belarus) with high extent of pollution from 1986 to 2000 the indicator of the birth rate has decreased with 17.2 to 9.7, life expectancy has decreased 72 up to 67.6 years, mortality with 9.2 to 14.8 has increased. In Belarus mortality is twice more than the birth rate.

In 15-20 years after the atomic bombing of the Japanese cities reduction of life expectancy in affected regions was also observed.

The expressed aggravation of a number of chronic diseases, a part from which is caused by the direct influence of radiation, is noted. The indicator of days of disability and transition to disability has considerably grown.

Among various diseases, the growth of prevalence of a syndrome of chronic fatigue is noted. His emergence is connected with the impact of radiation on nervous and endocrine systems.

At adult's diseases of cardiovascular, nervous and digestive systems, sense organs belong to the main reasons for the increased primary disability. The listed indicators it is expressed have worsened at liquidators of consequences of the accident at the age of 40-45 years.

At children, the prevalence of many diseases, especially nervous, cardiovascular systems, bones, and cancer of various bodies, in particular, of the thyroid gland has increased.

Studying of the influence of radiation on population is of great interest. Experts consider that radiation of bigger number of people small doses is equivalent to radiation. The changes in the specified indicators in many respects was defined by the extent of radioactive pollution. So, in the Gomel region (Belarus) with high extent of pollution from 1986 to 2000 the indicator of birth rate has decreased with 17.2 to 9.7, life expectancy has decreased 72 up to 67.6 years, mortality with 9.2 to 14.8 has increased. In Belarus mortality is twice more than the birth rate.

In 15-20 years after the atomic bombing of the Japanese cities reduction of life expectancy in affected regions was also observed.

The expressed aggravation of a number of chronic diseases, a part from which is caused by the direct influence of radiation, is noted. The indicator of days of disability and transition to disability has considerably grown.

Among various diseases, the growth of prevalence of a syndrome of chronic fatigue is noted. His emergence is connected with the impact of radiation on nervous and endocrine systems.

At adult's diseases of cardiovascular, nervous and digestive systems, sense organs belong to the main reasons for the increased primary disability. The listed indicators it is expressed have worsened at liquidators of consequences of the accident at the age of 40-45 years.

At children, the prevalence of many diseases, especially nervous, cardiovascular systems, bones, and cancer of various bodies, in particular, of the thyroid gland has increased.

Studying of the influence of radiation on population is of great interest. Experts consider that radiation of a big number of people small doses is equivalent to radiation of big number of people high doses. It is calculated that the genetic risk for 100 people who have received a dose of 0.01 Stars is equivalent, from the point of view of the defeat of population, to risk for 10 people who have received a dose of 0.1 Sv and to the risk for one person who has received a dose of 1.0 Stars.

Therefore, many violations of the health of the person at radioactive effects differ from all other accidents and the fact that they genetic, i.e. pass from father to son and extend in population.

For systematization of all variety of influence of radiation on the person system and organ, the approach is chosen us. A great number of people high doses. It is calculated that the genetic risk for 100 people who have received a dose of 0.01 Stars is equivalent, from the point of view of the defeat of population, to risk for 10 people who have received a dose of 0.1 Sv and to the risk for one person who has received a dose of 1.0 Stars.

Therefore, many violations of the health of the person at radioactive effects differ from all other accidents and the fact that they genetic, i.e. pass from father to son and extend in population.

Haematogenic system

The main anatomic substratum of the haematogenic system is marrow and a product of his activity – (peripheral) blood circulating in vessels.

At a dawn of radiation biology and radiation medicine, the high sensitivity of the haematogenic system to influence of radiation is noted. It can become the reason of her serious illness. Therefore blood less often –bone marrow, become an object of scientists after radioactive effects (accidents, radio treatment, professional activity, etc.).

Soon after the accident on the CNPP, a number of quantitative and high-quality changes of cellular composition of blood (erythrocytes, leukocytes, and platelets) are found in adults and children. Some changes were passing and were assessed positively as the compensatory reaction of an organism to radiation.

In the next years changes of erythrocytes and a ratio of kinds of leukocytes and also decrease in the level of hemoglobin quite often were found in children. 10 years later in marrow changes came to light (a dysphasia, violation of blood supply and others).

The correct assessment of the found changes is possible only on condition of further observations. The heavy course of beam aplastic anemia is noted. The next 20-30 years growth of incidence of the haematogenic system, in particular, is predicted by leukemia.

Immune system

According to modern representations, the role of immune system includes protection of an organism against many external (an infection, chemicals, radiation others) and internal factors (a cage of various bodies, etc.). It is the first line of protection of an organism against adverse effects.

The role of immune system comes down to the protection of an organism against diseases.

A specified function is carried out by special bodies. Treat them: red marrow, a spleen, lymph nodes and lymphatic vessels, thymus special functions in a small intestine (Peyerov plaque), tonsils, adenoids, and an appendix.

"Executive" function of immune processes is carried out by white blood cages, the called lymphocytes. Two main types of lymphocytes – T- lymphocytes, and B-lymphocytes, are known. Their quantity in the blood course is about one trillion. Besides, many substances, the called mediators (transmitters) participate in mechanisms of immune processes.

It is known that the bodies relating to the immune system are very sensitive to radiation. Changes occur at radiation in a dose more than 100 Rem (0.1 Gr).

The radioactive effects on the immune system can cause three main types of violations - an immunodeficiency, an allergy, and autoimmune reactions. The immunodeficiency reduces resilience to infectious diseases, including very heavy (for example, AIDS). As a result, their prevalence increases. There was a new concept designated "the Chernobyl AIDS." The specified violation is observed quite often and is fraught with grave consequences.

The allergy defines mechanisms of emergence of a number of diseases (bronchial asthma, skin diseases, etc.). Autoimmune violations are the cornerstone of many diseases (system a wolf cub, insulin-dependent diabetes, etc.).

After radioactive effects, there are difficult violations of humoral and cellular immunity. Various changes of the immune system were defined in the first days after the accident, remain so far and will remain for a long time.

Much attention is paid to cytogenetic changes at the liquidators and persons living in the polluted regions. 10 years later after the accident, the cytogenetic changes which are directly correlating with a radiation dose have been found.

The immune system participates in mechanisms of the majority of the diseases caused by radioactive effects (cancer, genetic and endocrine diseases, damage of internals and many others).

Nervous system

Consists of three interconnected section – central, peripheral and vegetative. The substratum of the central nervous system is the brain, peripheral – nervous trunks, vegetative - nerve ganglia.

The main function of the central nervous system this implementation of all mental processes (thinking, memory and many other things).

The peripheral nervous system includes the nervous trunks departing from a brain and their branching in various parts of the body. Her function – transfer of nervous impulses from the center (brain) on the periphery and back.

The function of the autonomic nervous system includes regulation of the activity of all internals (heart, kidneys, a liver, etc.).

According to laws of classical radiobiology, the central nervous system belongs to bodies' resistant (insensitive) to radioactive effects. From the subsequent information, it will be obvious that the specified postulate should be reconsidered.

Various violations of mentality have been noted in early terms after the accident on the CNPP. They were tried to be treated as displays of a radio phobia (fear of radiation). Such opinion kind of emphasized lightness of the appeared symptoms. During the same period, there were signs of the defeat of various systems of an organism which tried to explain with violations of the autonomic nervous system. Therefore "Asthenic syndrome", "Vegetative dystonia", etc. became the most popular diagnoses.

Such irresponsible diagnosis was made also to heavy in patients. It is necessary to remember that heavy patients had also the second – the true diagnosis of radiation injuries. These data were stored in safes. The similar treatment was necessary allied and to the Republican governments for the purpose of concealment of the weight of medical consequences of the accident. For victims, the specified irresponsible activity has turned back untimely diagnostics, treatment and secondary prevention of serious violations and diseases.

Radiation violations of the peripheral nervous system aren't established yet.

Long-term, today already long-term observations, have allowed estimating correctly violations and diseases central and the autonomic nervous system. Multidimensional scientists give the grounds to consider that damage of a brain is caused both by the direct influence of radiation, and various violations arising in an organism of the irradiated person (a metabolism, immune system, etc.).

In favor of direct impact of radiation on a brain accumulation in him of some isotopes testifies. Special researches of tissue of brain have found changes in various cages and violations of biochemical processes.

The observations which for the first time have established distinctions of damage of a brain at the high single and repeatedly influencing small doses of radiation are of great interest.

At radiation the following violations and diseases can be observed by high doses: a brain form of sharp radiation sickness, the chronic progressing radiation sickness of a brain, widespread or focal changes (encephalopathy, etc.), tumors of a brain and brain covers.

Radiation by small doses of radiation is shown by early and late changes of nervous system and also congenital defects at posterity.

In the USA the research of mental health of a big group of persons arrived from areas with the increased level of a radiation background after the accident to the CNPP is conducted. It is revealed that in 16 years after the accident, the fear of possible consequences remains. Their degree and frequency are in direct dependence on the distance to the radiation source. In other words, then it is less than a distance; it is heavier than a consequence.

Psychological influence Chernobyl, as well as similar accidents, differs from other accidents (an earthquake, floods, etc.) in what generates a situation of chronic threat for a long time (years, decades and more).

The opinion is fair that the sharpness of perception at residents of the former Soviet Union is caused also by many previous cataclysms (revolution, World War II, etc.). The important role was played by the criminal policy of the government which is coming down to false information for the purpose of reduction of the true sizes and possible consequences of the accident. People ceased to trust.

They found changes in the central nervous system reveal staging and depend not only on a radiation dose. They have a certain communication with age and a sex of victims, the general condition of an organism and social conditions.

Violations of the central nervous system are extremely diverse. Treat the most frequent violations: decrease in memory and mental adaptation, violation of emotional reactions, depression, persuasive phobias.

Liquidators to who refer apathy, paranoid thinking, concern in abstract problems, thoroughness, and viscosity of speech production, the decrease in speech processes and purposeful activity have data on pathological change of the personality.

Seven years later after the accident, it is revealed what disorders of mental and emotional health with the largest force was shown among mothers with small children living in close proximity to nuclear power plant.

It is necessary to carry persons with similar violations in the past to risk group of changes of mental health. Doctors in Israel have come to these conclusions, observing the persons which have arrived from the polluted regions of the former Soviet Union. At children, the bad progress and predisposition to mental diseases are noted. Similar data after atomic bombing are submitted by the Japanese doctors.

The social and mental syndrome of victims which is characterized by behavioral deviations, a sleep disorder, and headaches are described.

Inspection of immigrants, early the polluted areas living in, has revealed a number of changes of the psycho-emotional sphere which are more expressed at women. Increase in prevalence of vascular and brain violations are noted. The arising diseases (encephalopathy, arterial hypertension, a sclerosis of vessels of a brain) differ in the progressing current and high risk of complications (a myocardial infarction, a stroke, etc.).

Similar violations met after atomic bombing and accidents on the NPP.

Special tests (electroencephalography) confirm organic character and stability of damage of many structures of a brain. Stability of damages is caused by the influence of long-living isotopes (Cesium, Strontium, etc.).

It is necessary to emphasize that the electroencephalography can reveal violations even before the emergence of subjective signs.

Among victims, especially liquidators of consequences of the accident, the number of suicides has increased.

Various violations of activity of the autonomic nervous system also differ in stability and are quite often combined with changes of the central nervous system. The most frequent manifestations of violations of activity of the autonomic nervous system are changes of warm and vascular regulation (violations of frequency and a rhythm of a pulse, fluctuation of blood arterial blood pressure, pain in the heart). Electroencephalogram indicators change.

Before the Chernobyl accident, such profound studying of mental health at the irradiated persons wasn't carried out. Quite naturally that many victims suffered and have a radio phobia. But she, fortunately, isn't the reason of serious violations of health. As a rule, phobias are observed also at various other accidents.

Cardiovascular system

Heart, the big arterial and venous vessels and their working departing from a concern to her. It is stated above that a number of symptoms of cardiovascular diseases can be connected with violations of the autonomic nervous system. At the same time, it is necessary to remember that similar signs can arise also at a serious illness of vessels and hearts. Increase in prevalence of diseases of cordial vessels and the heart attack of a myocardium. Mainly at persons to the 45th summer age is observed. Before the accident, these people were almost healthy. Refer his big sizes and fast development of heart failure to distinctive features of a myocardial infarction. At a microscopic research of a myocardium the heavy violations in vessels and a muscular cover designated by some authors the term "radiation myocarditis" are revealed. The sclerosis of a cardiac muscle at persons of young age is noted

It is revealed that the myocardium is the place of accumulation of radioactive strontium. It confirms their radiation origin.

Radiation is referred to risk factors of a formation of blood clots and atherosclerosis.

Violations of the cardiovascular system as it is stated above are the frequent reason of disability of persons of young age. The growth of a number of diseases of heart and vessels is predicted.

Respiratory system

Anatomic the respiratory system includes top airways (nose, throat, and trachea, large, average and small bronchial tubes) and lungs.

The changes of the respiratory system in many respects are defined by time which has passed after radiation. In the first weeks after the accident on the CNPP "The syndrome of radiation damage of the top airways" was observed. He was shown by pain and unpleasant feelings in a throat and a throat, dry, sometimes hoarse cough.

In the subsequent time, there was an atrophy (rare hyperplasia) .

The mucous membrane of the top airways. Slow inflammation was most often observed. In some cases at a microscopic research of a mucous membrane of airways, the changes qualified as a precancer are found.

Changes of the composition of electrolytes and minerals in a mucous membrane of airways also broke their function.

In airways appeared special, so-called melanin - the producing microbes. They reduce the resistance of the pulmonary tissue to various influences, including microbes. The heavier course of pneumonia remaining the long period of time after accident paid attention.

Within the first four years especially heavy, badly responding to treatment pneumonia designated as radiation was observed.

Long radiation of staff of the Mayik enterprise was the reason of severe damage to lungs – the plutonium pneumosclerosis diagnosed for 123 patients.

At liquidators and persons, it is long the polluted territories living on, the tendency to respiratory diseases and lungs is noted.

At a microscopic research, the changes of cells of airways qualified as a risk factor for lung cancer are revealed.

At adults and children the reduction of volume of breath, stagnation of blood which are a

risk factors for diseases of lungs is revealed. Among children, the frequency of respiratory allegros has twice increased, and that is atypical for children's age, the pneumosclerosis comes to light. The fibrosis of lungs caused by radiation was observed also in adults.

The close attention is required by the fact of the increase of tuberculosis of lungs and his severe forms (cavernous disseminated and also an allocation of bacilli) and death. According to the available prognosis, in the subsequent, the prevalence of tuberculosis will increase in all age periods.

Gastrointestinal system

Treat a gastrointestinal tract a stomach, thin and thick gut, liver, and pancreas.

Damage of digestive organs can be caused, both external and internal radiation (through food and water). Usually, the extent of damages increases at high doses.

The high radiation sensitivity of a mucous membrane of a digestive tract (stomach, intestines) is the cause of a number of diseases. Sharp ulcers and erosion are observed. In a mucous membrane of a stomach quite often there is atrophy, the formation of gastric juice and motility decrease. Throwing food from a stomach in a gullet – a so-called reflux syndrome. The listed changes accrued six years later after radiation. The course of stomach ulcer differs in the prevalence of a dyspeptic syndrome (nausea, an eructation, decrease in appetite), without pain. It leads to late diagnostics and, therefore – to untimely treatment.

Generally, the prevalence of diseases of a stomach at adults and children has increased by 1.5-2 times. Increase in prevalence of cholecystitis and the angiocholitis are revealed. In later terms after radiation, there are malfunction and structures of a liver. These changes are designated as radiation hepatitis. Its prognosis adverse.

It is noted that heavy changes of a liver are the risk factor of viral hepatitis A and B.

Endocrine (hormonal) system

Carry the bodies producing the special substances getting to blood to endocrine system – so-called hormones. From here name hormonal system.

The hormonal function is carried out by many bodies (thyroid gland, adrenal glands, pancreas, etc.). Hormones take part in the most various processes of life support of an organism, both in normal conditions and at various influences (in particular, radiation) and diseases.   The hormonal function of endocrine bodies interacts with certain sites of a brain (hypothalamus), hypophysis, etc.).

The radioactive effects of the Chernobyl accident were made with, or other influence practically on all functional structures of the endocrine system.

In an initial stage after the accident, the deviations increasing the resistance of an organism to radiation are found. Such compensatory reaction is estimated as a positive factor.

On the other hand, the   impact of radiation on the systems which aren't participating in adaptation process has caused violations of interaction between the central and peripheral links of regulatory processes. Heavy changes of activity of many systems of an organism have resulted. So, the decrease in synthesis of sex hormones at men has led to the violation of a formation of sperm and ejaculation. Reduction of level of antioxidants was the reason of the changes in sperm contributing to the development of congenital ugliness in descendants. At women often there comes violation of a menstrual cycle.

Treats earlier a little-known disease radiation (autoimmune) thyroiditis. Especially often he was observed in the first weeks and months after the accident on the CNPP. Separate cases were observed also in the next years, mainly at the persons continuing accommodation in the polluted places. Believe that the risk of development of the earlier a little – known disease radiation (autoimmune) thyroiditis. Especially often he was observed in the first weeks and months after the accident on the CNPP. Separate cases were observed also in the next years, mainly at the persons continuing accommodation in the polluted places. Believe that the risk of development of the thyroid. It will remain for many years.

Radiation thyroiditis it is shown by decrease or increase in function of gland and increase in her size.

At decrease in function (hypothyroidism) in patients the general weakness, hypostases on a face and legs, decrease in body the temperature, apathy disturbs. Increase in function (hypothyroidism) is characterized by tachycardia, warm interruptions, weight loss, perspiration, temperature increase of a body, irritability. It will remain for many years.

Radiation tireoid It is shown by decrease or increase in function of gland and increase in her size. At decrease in function (hypothyroidism) in patients the general weakness, hypostases on a face and legs, decrease in body temperature, apathy disturbs. Increase in function (hyperthyroidism) is characterized by tachycardia, warm interruptions, weight loss, perspiration, the temperature increase of a body, irritability.

According to experts of Israel, 43% of the children repatriated from the polluted regions have those or other dysfunction of thyroid gland. Radiation thyroid it treats risk factors of cancer of thyroid gland. Experts believe that in the next years they increase of various violations of thyroid gland will remain.

The Japanese scientists have noted the increase in cases of a hypothyroidism at posterity of the persons which have undergone radiation influence as a result of the atomic bombing. Changes for the endocrine system are registered at adults and children. It is revealed what leads violations of some functions of endocrine system at children to lag of physical development and disharmony.

As a result of the decrease in function of a pancreas, the prevalence of a severe form of diabetes has considerably increased.

The incidence of bodies of the endocrine system in various regions has increased at 9-28 times. The presented materials it is possible is generalized to qualify as violations of a hormonal homeostasis.

Kidneys and uric ways

Kidneys not only body of formations of urine. They take great interest in removal from an organism of the exogenous, i.e. entered from the outside substances (for example, some isotopes). Besides, kidneys participate in the regulation of blood arterial blood pressure, blood formation, some types of a metabolism. The activity of kidneys is interconnected with other systems of an organism (cardiovascular, endocrine, etc.).

Uric ways are urethras, bladder, and urethra.

Due to list, it becomes clear that radiation can't but exert the impact on the activity of kidneys and uric ways. The damaging impact of radiation on kidneys is possible as at internal and external radiation.

Long ago the special type of inflammation of kidneys – the radiation nephrite developing after treatments by X-rays of bodies of a small pelvis has been described. He differs from radiation nephrite after the accident on the CNPP in the fact that arose through a wide interval of time (15-20 years) after radiation. The specified feature belongs also to nephrosclerosis (growth in kidneys of connecting fabric) with subsequent hypertension and a renal failure.

Radiation nephrite after the accident on the CNPP developed in – earlier terms after the accident (in 5-10 years). The listed diseases a long time can proceed without any symptoms. Therefore they are subject to active identification (the analysis of urine and biochemical blood test, measurement of arterial blood pressure and others).

The violations of exchange processes arising under the influence of radiation, in particular, peroxide oxidation of lipids of membranes of cells of renal tubules, consider as a risk factor of interstitial nephrite including at children. Besides, these processes promote the formation of stones in kidneys and uric ways. Allocation by kidneys of some radioactive materials as it is specified causes development of cancer of uric ways.

Metabolism

Long-term observations over affected persons have allowed revealing a number of violations of main types of a metabolism: fatty, protein, carbohydrate, electrolytic, minerals, and vitamins. Similar data are obtained in experiments on animals.

Especially expressed there were violations of exchange **of fats.** Activation of so-called oxidation is observed. It creates prerequisites for developing a serious illness among which atherosclerosis and cancer. The listed violations are found both in adults and in children. It is noted that violations of peroxide oxidation reveal a certain communication with age a floor and heredity. Violations of exchange of fats it is qualified as the risk factor of premature of an organism and atherosclerosis. In four years after the accident the specified violations decreased but didn't reach norm.

Changes in exchange of **proteins** are resulted by the deficiency of sulfur-containing amino acids leading to many violations in an organism. Increase in protein content in blood and change of a ratio of his fractions is observed.

Violation of regulation of exchange of **carbohydrates**, in particular, insufficient production of insulin of a pancreas, leads to increase in level of the glucose in the blood. Everything listed exerts a negative impact on other types of a metabolism and activity of some cells.

Changes of exchange of **electrolytes** (potassium, sodium, calcium, etc.) it is shown by an imbalance – decrease in content in plasma of blood and increase in erythrocytes. The listed changes differ in stability and partly arise in connection with violations of some bodies of hormonal system (for example, adrenal glands).

At the irradiated persons exchange of minerals suffers (iron. zinc, cobalt, etc.). Almost four times the content of iodine in blood increased.

Changes in exchange of vitamins come down, the main thing primarily, to deficiency of A, C, E vitamins and beta- carotene.

Hearing and vestibular device.

The decrease in hearing is described. It sometimes comes to light only by means of the special device – the audiometer. Researchers from the European countries consider that the radioactive effects belong to the new reasons of decrease in hearing in 21 centuries.

It is established that radiation steadily negatively influences all sections of the hearing aid. A hearing disorder has found certain confirmation on electroencephalograms.

Along with a hearing disorder, or in itself, often there are violations of activity of a vestibular mechanism which is shown sometimes heavy dizziness's. Similar violations are noted at 70-79% of liquidators.

Organ of vision.

Victims can have radiation glaucoma (increase in intraocular pressure) and cataract (cataract).

The crystalline lens is most sensitive to radiation. Radiation cataracts differ from age in the fact that the cataract begins with the periphery, but not with the center. Their heavier current is noted. After atomic bombing development of cataracts was observed in various terms – from three months to 10 years. They arose more often at the persons which are closer to explosion epicenter.

Diseases of eyes, including cataract, more often happen at the persons who have undergone radioactive effects at children's and youthful age. For example, in Austria where the raising of a radiation background after the accident on the CNPP was low, congenital diseases of eyes at children whose embryonic development was the share of the period of increase in a radiation background were observed.

At early stages, after an accident, most of the liquidators had changes of the forward part of an eye of eyes reminding ultra-violet burns.

In the next years, the growth of age cataracts and their emergence at relatively young age is noted. The same regularity was observed in changes of a mesh cover of an eye and her vessels. The increase of cases of a cataract at astronauts is revealed.

Anomalies of refraction, the disease of a mesh cover, an atrophy of a plaintive sack, a symptom of "dry eyes" are observed. At children, the changes of the back part of eyes similar to that at the persons who have endured atomic bombing came to light. Besides, cataracts caused by immune violations are described.

Skin and its appendages

Skin. Long influence of small doses of radiation brings to early to unknown changes. They, mainly, settle down on hands, feet, the lower third of shins, i.e. on open surfaces. In rare instances, changes cover 30% of a surface of a body. Formation of superficial ulcers, the emergence of a gematolimfoma is characteristic expansion of vessels, consolidation, and growth of skin elements (keratosis). Violations of exchange of pigments which are estimated tumors are found.

Hypodermic radiation fibrosis and especially - keloid growths belongs to very heavy changes of skin as risk factor developing of a malignant tumor

Keloid excrescences and hypodermic fibrosis in place of burns of skin to the very heavy changes of skin hypodermic radiation fibrosis behaves and especially is keloid excrescences. As a risk of origin of the malignant tumor factor.

Nails. In the area of the nail bed observed hemorrhage, increased pigmentation (discoloration), and subcutaneous fibrosis. These changes are usually combined with radiation damage to other organs.

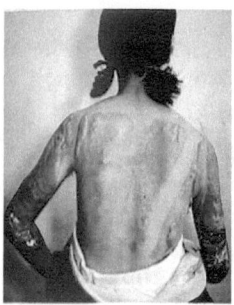

Hypodermic radiation fibrosis

Hair. Discovered a mutation, i.e. change the properties of hair an expression which can be baldness. In addition, thinning hair, changes their normal color. A few years later in the hair of affected individuals found high levels of radioactive plutonium.

Teeth. In tooth enamel accumulates radioactive strontium. Its level is especially high in men, possibly linked to their greater participation in liquidation of consequences of the accident. As stated above, American authors proposed to determine the contents of strontium in the teeth to assess the level of exposure. There is more frequent tooth decay, destruction of the gums (periodontal disease) and the slow growth of teeth. There are data on the increased incidence of teeth in children.

Bone system.
For liquidators, a syndrome osteopening the basic sign of that is a pain in a lumbar area is described. At the special x-rayed research found out dilution of bone fabric, sometimes small breaks in the bodies of vertebrae, osteochondrosis, and spondylosis. In bones more often tubular (hands. feet), observed sedimentations of radioactive cesium. The increase of cases of malignant tumors (sarcoma) is marked, especially for children

Present report about the origin of malignant tumors sarcoma of bones of unusual localization.

There are messages about developing of malignant tumors (sarcoma of bones) of unusual localization.

The presented materials convincingly demonstrate the variety of radiation violations and diseases at the long influence of small doses of radiation. Along with a variety of radiation damages, there is their direct link with a dose and duration of radiation and also other factors (age, quality of treatment, etc.)

TREATMENT OF RADIATION DAMAGES AND DISEASES

Today in the market there is no effective medicine which can cope with displays of radiation sickness, but several similar means are at a development stage. The main part of medicines which are used for radiation sickness today is effective only when to enter within 24 hours after radiation exposure.

Treatment, as well as prevention of radiation damages, and diseases are based on various classifications. Their combination promotes progress.

Classifications and mechanisms of action of protectors of radiation contain a chemical structure of medicines or duration of their action.

Radio protectors of short-term action.

A. Containing sulfur radio protectors happen four groups.

1/mediate influence the excited molecules, at the time of influence of radiation their physical state by restoration of an electron layer.

2/. Temporarily, active molecules of a biological substratum reversibly oppress "protecting" them from defeat.

3/.Inactivate the formed fat acid radicals at a stage of formation of hydro peroxides, then block chain reactions and significantly reduce the number of radio toxins in a lymph.

4/. Connect bivalent metals – oxidation catalysts that promote break of reactions of peroxide oxidation;

5/. Strengthen drainage function the reducing intoxication of the lymphatic system that is shown in the increase in a location of lymph.

B. Biogenous amines. The mechanism of the protection of medicines of this group is connected with an angiospasm and circulator changes of blood supply in radio sensitive bodies and fabrics. Therefore the hypoxia defining protection of these fabrics develops. Biogenous amines reduce the frequency of chromosomal aberration and by that risk of formation of tumors.

Distinguish two types of biogenous amines.

1/ Medicines with estrogenic activity. The condition of an increased estrogen which defines increased in resistance of phospholipids of membranes to processes of free radical oxidation and increased antioxidant activity of lymph, in general, is the cornerstone of the mechanism of protective action. Sharing of a Tsisterin and DES provides a more expressed effect in comparison with that which develops at the application of these radio protectors separately.

2/ Polysaccharides, nucleonic acids, and synthetic polymers. Them biological mechanisms, with ability: to connect the cornerstone high-molecular connections, to stimulate synthesis of nucleonic acids; to settle cells of marrow, young, capable to reproduction, in the irradiated organism; to form new and to activate the remained blood formation centers by fixing of cells of marrow in the struck haematogenic tissue.

The history of treatment of radiation damages contains over 100 years. For this period there were considerable changes in many fields of science and medical practice. Indications and method of use (doses, a way of introduction, etc.) are defined individually. Only in the last 50 years treatment of radiation damages, as well as other aspects of radio medicine, came to the state level. Various profile institutions with an exact regulation of their functions are created and are created.

Already today the certain success in the treatment of radiation injuries based on achievements of many sciences is made. Nevertheless, treatment has to be improved. Treatment of radiation defeats and diseases is the multidimensional and very complex problem. In the specified problem, it is possible to allocate two main directions.

The first direction includes the use of methods of neutralization of radioactive materials.

The second direction is a treatment of consequences of radioactive effects and the diseases caused by him.

It is made obviously that the maintenance of the specified directions is defined by all characteristics of radioactive effects (a dose, ways of penetration of isotope and many other things) and also country socio-economic indexes.

Neutralization of radioactive materials. The central place of neutralization of radioactive materials and the decrease in their negative impact on an organism has defined the emergence of the new section of radio medicine – a radio patronage. The substances neutralizing isotopes and their impact on an organism are called protectors of radiation.

The radio patronage has gained especially intensive development in the last 50 years. For this period over 400 radio protectors have been studied. The few of them were suitable for the person. Protectors of radiation shouldn't be toxic, i.e. cause serious collateral effects. Therefore it is necessary to refuse many substances. Scientist were conducted and conducted both in an experiment and at the people who have undergone radioactive effects. Such scientists on certain programs are conducted by large experts in specialized medical institutions.

For the first time in 1949 as protector of radiation cyanide of sodium and cysteine have been used. From 60th years of the 20th century, the increasing attention is drawn by so - called biological protectors of radiation.

Substances of natural origin with various properties concern them (adaptogens, antioxidant, blood stimulants, immune stimulant, antimutogenic and others).

The listed properties of radio protectors are directed to many pathological processes arising from radioactive effects.

During the Chernobyl accident, protectors of radiation weren't used. Only the pilots of helicopters who were taking part in accident elimination received medicine of domestic production Indralin. It isn't possible to judge its efficiency because of a lack of publications on this subject.

The history of the creation protectors of radiation in many respects is defined by the level of development of many divisions of science (radio physics, radio immunologists, medical pharmacologist, and many others).

Till 40th years of the left century studying of radio, protectors were carried out, mainly, in experiments on animals.

After atomic bombing of the Japanese cities, protectors of radiation began to be used at sick people. The Chernobyl accident was a new push for expansion and deepening of development of protectors of radiation.

The intensification of such researches is also promoted by the probability of application by terrorists of "a dirty bomb".

Certain positive results are as a result received. Methods of their systematization are offered.

Allocate three groups of the substances and medications having radio tire-tread effect.

1. Dekorporant - the substances accelerating removal from an organism of radio nuclides. All of them are well dissolve in water and are quickly allocated from an organism, generally with urine. Dekorporant is applied through a mouth or are entered intravenously. These substances are effective in early terms later on the decrease of radio nuclides in an organism. In high doses, they are especially effective for the isotopes which are late in bones (potassium, strontium, etc.). The majority of dekorporant is a little specific and can bring out of an organism useful substances (vitamins, minerals, etc.). Therefore duration of their application shouldn't exceed several days.

Recently the attention is drawn to dekorporant of photogenes. Their positive feature is the small toxicity in this connection their application is possible for a long time.

2. Enterosorbents, i.e. the substances capable to absorbs orbs) radio nuclides and to bring them out of an organism with a stake and urine. Each sorbent is specific, i.e. absorbs one-two substances. Therefore works on creation of complex enterosorbents are carried out. They are used not only in the medical and preventive purposes. Very important the fact that enterosorbents, even at prolonged use, don't remove substances, useful to an organism.

Creation of the medicines and nutritional supplements not only the radio nuclides blocking absorption and promoting their removal from an organism is perspective. Stone and absorbent carbon, high cleaning cellulose, silicon, organic gelatin medicines belong to the most widespread enterosorbents. It is noted that removal of radioactive materials from an organism is promoted by a sauna, a jacuzzi, and other hydro procedures. A similar impact is exerted by the Ukrainian mineral waters ("Naftusya", "Obolon" and others.).

3. Modifiers of radiation injury. The factors of the physical and chemical nature which are positively changing reaction of an organism to radiation concern them. Temperature, light, electromagnetic radiations belong to physical modifiers. The especially powerful impact is made by ultraviolet and electromagnetic fields of ultrahigh frequency. Modifiers exert impact at the different levels of the biological organization – from molecular to organism. It is necessary to carry the method which is recently offered by the American scientists to modifiers of beam influence hyperbaric oxygenation. Proceeding from modern ideas of mechanisms of radioactive effects, her use is quite expedient. Authors have a certain positive experience of the application of the specified method at radiation defeats at the patients receiving very intensive radiotherapy.

As the treatment-and-prophylactic means blocking accumulation in a thyroid gland of radioactive iodine use iodide potassium. The drug begins to be taken to an entrance to the polluted zone on one tablet (0.125 g) daily, within 3-7 days.

Reception of one tablet (130 mg) a day within two weeks is recommended to adults. To children, the dose decreases according to age. Medical effect of medicine comes down to the fact that accumulation of radioactive iodine in thyroid gland decreases.

The efficiency of treatment increases only at its application in early terms after radiation. Intake of iodide potassium for the purpose of prevention of damage to the thyroid gland is incomparably more effective. The mechanism of action of chemical protectors of radiation includes the change of the cellular reactions, therefore, resilience to radio nuclides and also the functional activity of cellular membranes increases. Very important the fact that the specified positive processes happen in the regulating and operating systems (central nervous, hyperphysical adrenaline, etc.). Protectors of radiation received from theorizing, beta- carotene, aromatic connections and other substances. Depending on action duration, they are used as for the treatment, and prevention.

The threat of radiation terrorism intensified development at the high methodical level of new ways of treatment. Similar researches are in the USA under systematic control of public institutions. Now they include receiving the medicines making an impact on radiation damages at the molecular, cellular and fabric levels. New classes of substances for the treatment of radiation defeats are in recent years created, their systematization is offered.

Distinguish four kinds' protectors of radiation.

1 / Medicines the containing izoflavine.

2/ Medicines reducing radioactive influence. They include a large number of names of the drugs exerting the positive impact on various mechanisms of radiation damage. Most of them are in a stage of a clinical check so far.

3/ Eliminators, i.e. the means promoting allocation from an organism of radio nuclides.

All these medicines are under certain codes so far and wait for the end of clinical tests. Recently in the USA three medicines – Ca-DTPA, Zn-DTPA and Prussian blue promoting allocation from an organism of plutonium, zirconium, and americium are allowed for the clinical application. They are shown only at internal radiation. Allocation of radioisotopes is carried out with a stake and urine.

Their use is contraindicated pregnant women and persons up to 18 years. Drugs are injected intravenously or intramuscularly. During treatment, the nutritional supplements containing zinc and magnesium are appointed. Side effects include nausea, vomiting, diarrhea, temperature increase of a body. After medicine cancellation side effects decrease. The efficiency of treatment decides on the basis of removal of isotopes on urine: their increase is regarded positively. Treatment is carried out only in the conditions of the hospital.

Prussian blue is applied through a mouth on 500 mg in capsules of times a day, or by inhalations. Duration of reception is defined depending on a radiation dose. Medicine adsorbs the isotopes which are in intestines which then are removed with a stake. Pregnant and small children can appoint medicine. Treatment is carried out only under observation of the doctor.

4/ The blocking radioactive materials (for example, iodide potassium).

Believe that for more effective treatment it is necessary to use previously biological methods of radiation damages (definition of aberration of lymphocytes in peripheral blood). Radio protectors of radiation are applied at radiotherapy and are shown to the persons working with radioactive materials.

According to leading experts, protectors of radiation have to correspond to the following qualities:

1) prevent sharp and chronic radiation defeats;

2) to be accepted through a mouth;

3) to extend quickly in an organism;

4) to be chemically steady;

5) not to render side effects.

As a radio protector, the Russian authors recommend to use the medicine Aktovegin. It can be applied in tablets, intravenously and in ointments. Medicine improves Microcirculation which violations happen at radiation injuries.

Experts warn that sometimes are issued and medicines which radio projector properties are very doubtful are widely advertised.

It is known that in the treatment of radiation injuries antibiotics figure prominently. It has turned out that the antibacterial treatment which is often shown to such patients represents a number of features. In this regard are offered and new antibacterial medicines are used. Their application has to be combined with antifungal medicines.

Focal encephalopathy is the indication to the systematic application of a Cerebrolysinum, the glutamine acid, peacetime and other means improving metabolism of a brain. At local necroses of a brain dehydration therapy, elimination of the central violations of breath, blood circulation and other disorders of stem functions is shown. In case of an exit from a coma hold rehabilitation events.

For improvement of function of bone marrow, along with early the applied medicines (vitamins, iron, etc.), offer new. Various options of erythropoietin's – the substances strengthening activity of marrow concern them.

Indications for transfusion of whole blood and its components elements of blood erythrocytes, leukocytes, platelets), and plasma are regulated. The complex of the actions directed to the treatment of damages of a gastrointestinal tract is developed.

Due to the Chernobyl accident, the extreme relevance was acquired by the creation of medicines for prolonged use in the conditions of the influence of small doses protector of radiation.

To dress with synthetic protectors substances of natural origin are revealed. Many of these substances, according to literature, possess the expressed adaptogenny action, i.e. promote adaptability of an organism, in particular, to radiation.

The use of the so-called functional products rendering favorable effect on a number of violations is recommended. Carry to them: grain mixes, chicory, food fibers, sea products.

Among biological protector's medicines of a ginseng, an eleterokok, beta - carotene and a media hydrolyzed (MIGI-K) are most fully studied. Radio protective properties of medicines of a dog rose, mountain ash, ordinary silverweed, yarrow, sea cabbage are noted. A positive feature of biological protectors of radiation is the lack of collateral effects and a possibility of longer reception through a mouth.

In-depth studies have installed some mechanisms of action of many biological radio protectors. It allows using medicines depending on the nature of damage.

So, beta - carotene and a plantain reduce the damaging action of radioisotopes by membranes of cages.

Medicine from mussels stimulates immunity has to stop bleeding and stimulates recovery processes.

The Ukrainian authors have offered complexes of gains, plants decoctions.

Leaves birches, green stalks of oats sowing, leaves of walnut and yarrow in the ratio 1:2:2:1. One tablespoon of the mix is filled in with a glass of the boiling water. To accept on a half of a glass 3-5 times a day in 30-40 minutes prior to food.

Hips-15 gr., hawthorn fruits – two-three gr, sea-buckthorn fruits – 15 gr. To fill in a mix with a glass of water, to bring to boiling, to insist 1.5 hours. To apply 150-200 ml of 4-5 times a day.

The mechanism of action of the offered plants is up to the end not studied. However, the made observations testify in favor of expediency of their use. A number of biological protectors of radiation have the good prospect of introduction thanks to sufficient raw material resources. They are used as in itself, and in the form of nutritional supplements and dietary products (bakery products, drinks, etc.).

Use of that or another protector of radiation depends on many factors of radiation defeat. Therefore in sharp and hard cases, intravenous means are shown. At long influence of small doses including at radiation treatment, the nontoxic substances entered through a mouth are recommended.

The role of so-called free radicals in development of radiation damage defines expediency of wide use of antioxidants.

Besides drugs, consumption of food rich in antioxidants, including vitamins is of great importance.

B-1: blackcurrant, parsley, a wild strawberry, a citrus, tomatoes, potatoes, bread from coarse flour, wheat sprouts, rice, honey, nuts, meat, milk, bean, corn.

B-2: leaves of vegetables and bushes, apples, the sprouted wheat, milk, eggs, a liver, eggs, milk.

B-5: cellulose of plants, meat products, liver.

B-6: yolks of eggs, the cabbage, grain cereals which have sprouted wheat, liver, kidneys.

B-9: soybeans, wheat sprouts, peas, lentil,

 cabbage, tomatoes, spinach, liver, meat, oats, nuts, bread, cheese, bananas, oranges

B-12: egg yolk, milk, liver, kidneys, parsley, apricots.

A: liver, fish, egg yolks, sour cream, milk, carrots, beet,* D: liver, creamy and vegetable oil, milk, cod-liver oil.

E: sunflower oil, sunflower seeds, almonds, germs of wheat and oats, peanut and olive oils, egg yolk, green peas, apples.

K: cabbage, spinach, root crops, fruit, liver, yeast.

The role of the vitamin C exerting the positive impact on many pathological processes is especially big bleeding, infection, etc. It is offered to appoint vitamin C depending on a concrete situation. At radioactive effects, the dose has to be increased to 5.0 g. / days. In case of heavy damages, the dose of vitamin C is recommended to be increased to 50.0 g. (!) in days and to enter him intravenously. The quoted author has also offered the combinations of natural nutritional supplements including below the given sets.

Vitamin C (1g. 3-4 times a day), vitamin E (400 International units 3 times a day), Q 10 coenzyme (100 mg 2-3 times a day).

- Glutathione (500 mg 2 times a day).
- Vitamin A (10 thousand international units)
- Beta carotene (25 thousand international units of times a day) zinc (30 mg of times a day).
- L-glutamine (3-10 g. a day).
- Bromelain (2500-500 mg between meals).

Treatment has to be carried out in a complex and step by step taking into account a number of circumstances (the sanitary and hygienic conditions, the characteristic of defeat conducting symptoms, etc.). For victims of radiation, special diets are developed. Increase in daily intake of protein by 15% and reduction of fats – for 30% is shown. The use of a large number of vegetables and fruit, seafood is shown (a sea cabbage, etc.). Food has to contain many fresh fruit and vegetables, bean and anti-inflammatory (cold water, nuts). It isn't recommended to use coffee, alcohol, sugar, spices.

By drawing up a diet it is necessary to remember that accumulation of isotopes in various foods unequal. So, cesium and strontium collect in milk, meat, potatoes, mushrooms, wild berries. To a dress with the reception of that or another protector of radiation, the victim has to be as soon as possible evacuated from the polluted area. Sanitary and hygienic processing irradiated is very important (a shower, rinsing mucous, change of linen and footwear).

Treatment of the acute radial illness (ARS) needs to be begun as soon as possible and to carry out in the conditions of the hospital, it is better specialized. Patients are located in the separate sterile, equipped with monitors chambers, under systematic observation of specially prepared medical personnel.

Treatment is carried out in a complex taking into account the main damages and the leading symptoms. Transfuse blood and its separate components (erythrocytes, leucocytes, thrombocytes), enter large amounts of liquid and electrolytes. Antibiotics, against fungal drugs, are used.

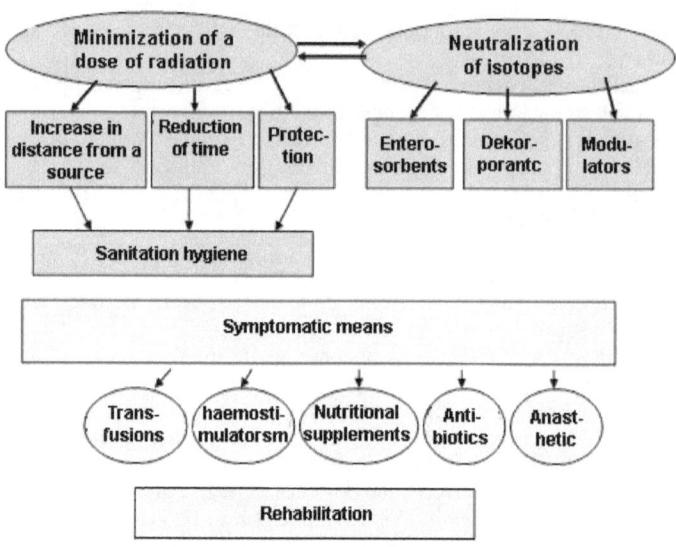

Scheme 3. Treatment of radioactive lesions

For stimulation of a hemopoiesis prescribe body height hormone, stem cells are entered. The symptomatic treatment is regularly prescribed (against an anemia, heart, and vascular drugs, etc.). By experience of the Chernobyl accident, bone marrow transplantation didn't justify the expected results. In hard cases not earlier 2-3 weeks after radiation, transplantation of marrow is shown

Also, its separate components (erythrocytes, leukocytes, platelets) transfuse blood; enter large amounts of liquid and electrolytes. Antibiotics, antifungal medicines are used. For stimulation of blood, formation appoints growth hormone, stem cells are entered. Symptomatic treatment is regularly appointed (against an anemia, heart, and vascular medicines, etc.). By experience of the Chernobyl accident, bone marrow transplantation hasn't justified the expected results. In hard cases not early 2-3 weeks after radiation, bone marrow transplantation is shown.

Use of laser radiation for improvement of activity of the immune system is offered. In case of developing of infectious diseases appoint antibiotics (Penicillin, Kanamycin, Rifadin). Antifungal medicines are at the same time applied. Anesthetics and antihistaminic medicines are usually used. The medicines improving function of a liver are applied.

Due to the damage of a mouth and digestive tract, there are great difficulties with food of victims. The fruit candy, the crushed cauliflower, the beaten egg whites, and cream are recommended.

Treatment of widespread and deep burns begins with removal of impractical fabrics. In the subsequent apply all known modern a burn methods of treatment.

Much more heavy burns on mucous membranes of a mouth, a throat and airways respond to treatment.

Recently success in the treatment of a radiation skin syndrome and hypodermic fibrosis is achieved. The technique depends on a damage rate. Treatment lasts from several weeks to several months. At patients with hypodermic radiation fibrosis with success, the medicine "Lipsod" applied within 12 weeks is used. Its side effects aren't found.

At stable keratoma surgical treatment - removal of the damaged sites with the subsequent change of healthy skin is carried out. The combination of derivatives of a metal xanthene's (PTX) to alpha- tocopherol is assessed positively. Scientists of new protectors of radiation continue.

The presented materials convincingly demonstrate that treatment of beam damages has to be individual, consecutive and complex.

PREVENTION OF RADIATION DAMAGES AND THEIR CONSEQUENCES

Physical, chemical and biological methods are applied to prevention of radiation injuries.

Refer application to chemical methods the pharmacological chemical of means. As free radicals, having high chemical activity, have the main striking effect, there was a thought of an opportunity to inactivate free radicals with the help the pharmacological kinetic of means. Such medicines have been created and have received the name protectors of radiation. Works of a number of authors (it has been established that connections which part thiol and amine groups have to be posses protective action. The first radio protecting has been so received. Scientists of domestic scientists have shown that under the influence of a Tsisteamin resistance to radiation increases for 30-50%. And in general, the protective effect of any medicine can be expressed such concept as the factor of reduction of a dose of FUD – the coefficient specifying in how many times "decreases" a dose under the influence protector of radiation.

In the standard understanding, primary prevention of radiation defeats is very limited. It is connected with the suddenness of their emergence. Nobody knows when there is an accident or the atomic weapons will be used.

For prevention of acute radiation defeats, it is necessary to observe the mode of radiation safety which includes a radiation survey, radiometric control, control of radiation of staff, protection of staff

In prevention of radiation defeats neutralization of the regions polluted by radio nuclides and deactivation of radiation, waste is fundamental. The huge significance is attached to protection and protection of all sources of radiation. Primary prevention of radiation defeats is possible only under some conditions of production and also in the patients receiving radiotherapy.

The main methods of primary prophylaxis of radioactive damages when using radio diagnosis and radiotherapy are:

- improvement of devices and technique;
- rising of efficiency of personal protection of the patient and service personnel;
- regular comprehensive observation (general inspection, blood test, etc.).

In certain cases, same protectors of radiation can be recommended. So, for example, intake of iodide potassium is periodically recommended to the persons living near nuclear power plants. Intake of iodide potassium for prophylaxis of diseases of a thyroid gland, in particular, cancer, is absolutely shown at once after explosion only at the existence of radioactive iodine. This information has to be brought to the attention of the population and the relevant institutions. The references on the preventive intake of iodide potassium approved in the USA by State department of health protection and Federal bureau on supervision of foodstuff and drugs are given below.

The dose of potassium iodide depends on age: adult (up to 40 years) take 1 tablet (130 mg) children 3 to 18 years – half tablet (65 mg), children from 1 month to 3 years – the fourth part of the tablet (32.5 mg). Newborn to 1 month appoint 16.25 mg of potassium iodide in the solution

The feeding women have to accept a dose for adults plus a dose for the child according to his age. Not all authors agree with the expediency of intake of iodide potassium after the 40-year age when the probability of development of radiation cancer sharply decreases.

Intake of iodide potassium is canceled after achievement of normal level in the environment of radioactive iodine. Iodide potassium is contraindicated to persons with diseases of a thyroid gland hyperthyroidism, a craw and others), hyper sensibility (allergy) to iodine and skin diseases. Tablets of iodide potassium in the USA can be got in any pharmaceutical institution.

Preventive effect of iodide potassium has been established by the Polish doctors at once after the Chernobyl accident. Due to the increase in a radiation background of iodine-131, over 10 million adult population and 7 million children have received a preventive dose of iodide potassium (on 15 mg a day). At the specified dose side, effects weren't observed, and the prevalence of cancer of thyroid gland has practically not increased.

Iodine prevention after the Chernobyl accident in the republics of the former Soviet Union has begun not at once. Medicines of iodide potassium have appeared insufficiently, and the technique of application wasn't well-known including to medics. Its "effect" has been reduced to the emergence of a bigger number of allergic reactions.

For prevention of radiation defeats, a number of new medicines are offered in recent years. Prevention of radiation injuries at especially dangerous internal infection is improved. The main actions at incorporation of the products of nuclear division (PND) have to be directed to the acceleration of their removal from an organism. It is reached by use of means and methods of removal from the digestive tract, airways, blood, and places of deposition.

For acceleration of removal of the digestive tract, it is used Adsobar, Foreseen, Polisharmyn, calcium entered inside.

For the purpose of binding and acceleration of their removal from respiratory organs carry out inhalations. of a Pentatcin.

Adsobar – sulfate barium with the developed adsorptive surface reduces at 10-30 times absorption from the digestive tract of isotopes of strontium (Sr89, 90) and barium (Ba140). Medicine is applied inside on 25 g on 200 ml of water daily, during the stay in the polluted zone.

Foreseen – a sorbent- the adjustable complexes with the structure of cells corresponding to the cesium atom size. Except cesium (Cs134, 137) connects radioisotopes of rubidium and tellurium. Medicine is applied inside on 1 g by 2-3 times a day within 15-20 days.

Polisurmin – antidote of contact action – the sorbent strengthening removal from an organism of the radionuclide of strontium. Medicine is applied on 4 g on 200 ml of water daily, during the stay in the polluted zone.

Pentatsin forces of covalent communications. At the same time, specific chemical properties of radionuclide are substantially lost. Interaction with proteins stops, their deposition in bodies, joints, and bones is broken. Medicine can be applied inside 50 ml of 5% of solution in the form of 5% of the solution on 0, 25-1, 5 g daily or every other day. On a course of 20 injections. It is inhalation – 10% solution on 0, 1-0, 2 g within 20-30 minutes.

Pharmaceutical establishment of Russia recommends using a set of medicines which is contained in the so-called preventive first-aid kit.

In the conditions of sudden radiation (accident on the NPP, use of "a dirty bomb"), it is necessary to use all actions directed to the reduction of a dose of radiation, it's the duration and possible violations. Therefore the huge value has the fast notification including indications on the place and intensity of pollution, composition of isotopes, addresses of the next points of deactivation and medical institutions. Except that, special services have to carry out transportation of patients, provided with water and food, first aid.

Pharmaceutical establishment of Russia recommends using a set of medicines which is contained in the so-called preventive first-aid kit.

There are biological methods of prevention of radiation defeats directed to the long increase in resistance of an organism an action of a radiation.

The first group – psychogeneses adaptogenny (eleuterococcus extract, magnolia vine tincture, ginseng tincture). They increase the resistance of an organism to many adverse factors, including to action of AI. Take the drugs of this group in 10 days prior to an exit to RZM, till 20-30 drops in 30 minutes prior to food daily. An optimum course – 20 days.

The second group – poly histaminic and vitamin acid complexes (Amitetravit, Tetrafolivit, Inosine). Take the drugs in 5 days prior to an entrance on RZM on 3 g, 2 times a day after a meal, daily. Optimum course 2 weeks.

The third group – metabolites – metabolism modifiers – medicine of amber acid –. The drug is taken during the entire period of stay on RZM, on 1 tablet dissolved in water 2 times a day to food.

The fourth group – antioxidants (Tocopherol, Pyridoxine, Inosine, Ascorbic acid). Apply during the entire period of stay on RZM according to usual schemes.

Physical, chemical and biological methods are applied to prevention of radiation injuries.

Refer application to chemical methods the pharmacokinetic of means. As free radicals, having high chemical activity, have the main striking effect, there was a thought of an opportunity to inactivate free radicals with the help the pharmacokinetic of means. Such medicines have been created and have received the name protectors of radiation.

By works of a number of authors, it has been established that connections which part timolovy and etymology groups have to possess protective action. The first radio protecting – this has been so received. Scientists of domestic scientists have shown that under the influence of a tsisteamin resistance to radiation increases for 30-50%. And in general, the protective effect of any medicine can express such concept as the factor of reduction of a dose (FRD). It is the coefficient specifying in how many times "decreases" a dose under the influence of a protector.

For reduction of the duration of radiation that has important value for its outcome, it is recommended to give, whenever possible, first aid out of the center of radioactive pollution.

The organization of protection against influences of radiation influence is in the USA at quite a high level and is constantly improved. Deal with this problem also in some other countries. The population, the environment in places of possible infection (nuclear power plants, burial of radioactive waste, etc.) is exposed to systematic radiation monitoring.

In some countries, including in the USA, radiation monitoring in transport, and first of all in aviation and also in places of a bigger congestion of people (the subway, stadiums, etc.) returns to normal.

Now the responsibility for the listed work in our country is conferred on the Centers for control and prevention of diseases, Centers for Disease Control and Prevention (CDC). They work in common with many institutions (regular and local department of health, power, transport, agriculture, Federal Bureau on quality control of medicines and foodstuff (Food and Drug Administration – FDA) Federal bureau of investigation (FBI)–and many others.

Instructions for prevention of radiation defeats are available in many countries, including in the countries of the former Soviet Union. Their quality not always satisfactory, and realization insufficient.

It is natural that in the prevention of radiation defeats the central place belongs to a prevention of realization of action of possible sources of infection. In case of radioactive danger the population and special services, first of all medical, are constantly informed.

The place of pollution, level, the composition of isotopes, etc. is reported. All mass media are for this purpose used radio, the TV, newspapers, e-mail, letters, leaflets, lectures, conversations. Use of telephone communication it possible.

In a stage of development, there is a device combining the cell phone and the dosimeter of radiation. Preparation and retraining of medical personnel and other services are regularly carried out.

Realization of this activity is under constant control of public service of all levels. For this purpose, huge allocations are allocated.

After radiation explosion, all preventive actions are carried out in a radius from 10 to 20 miles (16-32 km) depending on the force of explosion and level of radioactive materials.

One of the basic rules of reduction of a dose of radiation is an immediate evacuation of victims from a zone of radioactive infection and sanitary and hygienic actions (a shower, a hairstyle of hair, rinsing of a mouth, change of clothes and footwear, a bed and. etc.). It is very important not only for the victim and for people around since clothes and footwear can be the way of distribution of radiation.

For radiation reduction in the inhalation way, to an exit from a zone of radioactive infection, it is recommended to use masks, gas masks, handkerchiefs, etc. For personal prevention, it is necessary to leave as soon as possible the infected zone, to put on a mask. Having left a zone, it is necessary to take a shower, to replace clothes, footwear, bags, etc. The polluted things should be put in a cellophane (plastic) bag. The applied medicines are shifted in cellophane kites. It is strictly forbidden to use the water and food which are in a zone of radioactive pollution

If radioactive materials are outside of rooms, it is necessary to enter immediately the room, to carefully close doors, windows, window leaves, and external fans.

Due to various penetrability of radioactive beams in a human body, simple methods of protection are used.

- Alpha rays - thin material (for example, paper).
- Beta - rays - dense fabric.
- Scale - rays - lead.

Prevention of complications of radiation treatment, along with above the stated recommendations, includes careful control of the used doses and the systematic observation directed to identification of possible side effects (blood test, immunological monitoring, etc.)

The multiparameter models allowing to define a dose of radiation and to predict complications are used.

The staff of the large American university has offered the model allowing the predicting extent of radiation injury of lungs at radio treatment for each specific patient. For the purpose of prevention of radiation damages of a mouth, the American authors recommend medicine Binyamin.

The persons working in the conditions of technogenic radiation (nuclear power plants, submarines, etc.) are obliged to observe strictly and constantly regulated safety measures, to be exposed regularly to dissymmetry by the most optimum methods, to carry out the tests allowing to reveal initial radiation changes (abberation of leukocytes in peripheral blood and others).

Prevention of the defeats at radiation alarm. A possibility of the application by terrorists of "a dirty bomb" and other sources of radioactive effects predetermine need of development of the corresponding rules and recommendations.

It is emphasized that in case of the announcement of radiation alarm, the people who are in the room shouldn't go outside, and waits for orders. Nobody can tell how this time will long last. Therefore in each house, it is necessary to be prepared for a possible situation as much as possible. By an example of the State of Israel, for these purposes, the special room has to be allocated. Specially equipped room – a gas-proof shelter is provided in new buildings in each apartment.

How does have to be the placement on a case of the radiation attack - the so-called home shelter (Shelter in Your Home) is equipped?

It is the most expedient to use the room in a basement or it is central to the located room with the minimum quantity of windows. In the allocated place there has to be a stock of all of the most necessary. This quantity is defined by a number of inhabitants. It is necessary to provide the place for pets.

The recommended set includes:

- Food of long storage (tinned vegetables, fish and meat products, dried fruits, etc.).
- Water at the rate of one gallon a day on the person.
- Baby food.
- Change of linen, footwear, clothes depending on a season.
- Paper towels, plastic ware.
- Plastic bags.
- Bedding (blankets, pillows, etc.).
- Battery radio receiver.

- Medicines.
- Vitamin C in powder.
- Toilet accessories.
- Lamps on batteries.
- Phone or Cellular phone
- Spare points, lenses, and means for their processing.
- Trash begs .
- Food for animals.
- Medical first-aid kit.
- Soap.
- Books, games, etc. for entertainments.
- Credit cards and money.

Products, water, paper products, ware, etc. prepare at the rate on three days. Every six months stocks of food, water and medicines are updated. Before an entrance from the street to the shelter, it is necessary to replace clothes and footwear and to leave them outside in cellophane bags. It is necessary to switch off a fireplace, the conditioner, to put a plastic rug at doors, to densely close doors and windows. It is necessary to take away from the street of pets.

Radio has to be turned on all the time for obtaining constant information and instructions.

It is known that radiation has no color, a smell, and taste. It is possible to learn about radiation contamination (accident," a dirty bomb") only from official specialized institutions.

What needs to be done after radiation explosion?

The offered recommendations include following actions.

- Immediately on foot to leave the infected area.
- Not to panic.
- Not to use the public and individual transport since it can be polluted.
- To enter immediately the next room.
- As soon as possible to take off clothes and footwear, to place everything in a plastic bag.
- To take a shower, or to wipe a body with damp tampons.
- To use all media (phone, radio, the TV).
- To monitor further indications of responsible bodies.

For evacuation of preschool institutions, the schools located within the radius of 10 miles from the place of the explosion there are special rules. Instructions for evacuation medical, governmental and some other institutions, nursing homes are developed. Methods of secondary prevention of radioactive defeat are regularly developed and improved. So, in Ukraine they include:

- protection of the population and personnel against radioactive influence;
- closing of dangerous radiation objects
- the measures limiting distribution of radiation;
- complex monitoring of the population, employees, and environment.

All these actions unite under the heading "Minimization of Radiation Consequences". It is obvious that the specified perspective activity has to be improved, in due time and fully be carried out and be public.

Unfortunately, listed, for many reasons, it isn't always carried out in time and in full.

Careful and systematic control of radiation contamination of food and water is of great importance.

In Ukraine, it is established admissible the content of cesium and strontium in the main food. Carrying out such tests is possible only in special laboratories. Norms in Ukraine differ from Belarusian and Russian. Level of admissible doses isn't always represented adequately. According to the last messages, a number of amounts of isotopes in food much more exceed the established norms. It is noted that first of all, it belongs to products of the private sector where in fact, there is no control.

It is necessary to remember that accumulation of radioactive materials unequal in various food. In Ukraine, it is established admissible the content of cesium and strontium in the main food. Carrying out such tests is possible only in special laboratories. Norms in Ukraine differ from Belarusian and Russian. Level of admissible doses isn't always represented adequately. According to the last messages, the number of isotopes in food much more exceeds the established norms. It is noted that first of all, it belongs to products of the private sector where in fact, there is no control.

It is necessary to remember that accumulation of radioactive materials unequal in various food. So, for example, in the countries of the former Soviet Union for employees of radiation sources, admissible doses made 50 µSv in a year, for the population earlier – 5 µ/Sv. Since 1990 they are lowered respectively to 20 and 1 µ/Sv in a year.

All persons which have arrived in uncontaminated areas have to be subjected to all-clinical inspection with the assistance of a number of experts (the therapist, the neuropathology's, the endocrinologist, the allergist) and to undergo radiation control.

It is authentically established that late radiation consequences can arise in various terms after radiation. Their character in many respects depends on the received dose, duration of radiation, the name of isotope, specific features of the victim, external conditions

All persons which have undergone radioactive effects and also their descendants have to be under constant observation.

The methodology and technique of observation of victims are defined by a number of factors the Frequency of inspection and their volume are defined as quantitative (the size of a dose and its characteristic, terms after radiation, age), and results of the carried-out treatment, existence of the burdening circumstances.

For further observation over the victim, it is very important to know the size of the received radiation dose and also the name of the isotope. The specified data are available not for all victims in the former Soviet Union.

Use of that or another method in many respects depends on terms after radiation, the received dose, and properties of isotopes.

In this regard, it is necessary to apply all complex of the methods allowing establishing directly or indirectly a radiation dose.

A feature of radiation diseases is the long asymptomatic current. Therefore without fail it is necessary to reveal them actively.

Various techniques of the subsequent observation (medical examination) of victims of radioactive effects are described.

They have to be carried out depending on a number of factors.

1/ Doses of radiation and her characteristic and terms, age and floor.

2/ The carried-out treatment and its results.

3/ The burdening circumstances (social conditions, other diseases, smoking, alcoholism.

The persons which were injured from radiation need introspection which purpose is a timely diagnosis of possible diseases (for example, cancer of a thyroid gland, a disease of kidneys).

Great preventive influence on consequences of radiation is exerted by the periodic stay of victims in clean regions and the improving centers. Such centers are available in a number of the countries on various continents.

It is necessary to mark the great preventive importance of scientific educational work among all segments of the population (a grant, the brochure, a lecture, TV, and broadcasts). It is important for persons of all age and various social levels.

With fundamentals of radiation medicine, there has to be the acquaintance all. And it means to have ideas of radiation sources, methods of diagnostics, treatment, and prevention of radiation injuries.

The importance of an active position of the population for improvement of diagnostics, treatment, and prevention of radiation injuries can't be overestimated.

Scientific knowledge in these areas is systematically improved and it has to take root quickly not only into the activity of experts. All need to know about them. Questions not of knowledge of the population are subject to the immediate decision.

The scientific data on prevention of radiation injuries and their consequences received so far, promote the improvement of the prognosis.

CONCLUSION

The left century, start out new generations and memory even about the most destructive natural cataclysms, about the most an awful natural disasters is gradually erased. Yes, it is erased, except those cases when consequences of similar events are shown to this day. Therefore it at all not memoirs, but the present event.

Common sense as though has to prompt that if people were so ingenious to open, for example, X-rays and to split the atom, then they at the same time, exactly at the same time had to offer "old jackets for new monsters". However, it hasn't occurred. A number of scientists expected and were afraid of possible serious consequences of the specified opening.

As many great opening, X-rays, and splitting of the atom were ambiguous for mankind and nature.

On the one hand, opening the X-ray and obtaining atomic energy promoted and promote the progress of many sciences, including medical, to the rapid development of power and progress of a number of industrial productions.

On the other hand, atomic energy became a source of a lethal weapon which doesn't manage to be forbidden yet. Objects of nuclear power and other sources of radiation are unsafe (for example, radiation waste). It would seem, radiation accidents are the basis for sharp restriction and even the termination of use of radioactive materials. Such point of view is supported by some scientists and public organizations.

And if to look around, then it is visible that nobody stops dangerous flights by spaceships, to use cars, to float on huge sea liners because Titanic, Nakhimov, etc. has sunk.

Today it is already impossible to refuse the application of radiation for scientific research, in medicine, power, and many, many other areas. Therefore much attention is paid to the improvement of technology of receiving and use of radioactive materials and radiation protection.

In the last decades the in-depth studies devoted to influence on the person of natural radiation are conducted (space beams, uranium ores, etc.).

Receiving and use of artificial sources of radiation all the time is improved. A lot of things are made, but the danger isn't overcome yet. The damaging action of radiation on the person and nature, in fact, is known since her opening. For many years of intensive researches of experts of the various profile (physicists, biologists, doctors, radio genetics and many others). In a number of the countries, the huge actual material demonstrating the variety of influence of radiation on the person is saved up.

Radiation accidents differ from all others (an earthquake, a flood, the fire, etc.) in the fact that their consequences remain for many years and even centuries and exert the negative impact on the population. It is established that artificially created radioisotopes more and more widely used by the person are more aggressive also long-living than natural. And it fraught with more serious consequences.

After atomic bombing of the Japanese cities of Hiroshima and Nagasaki of a consequence have been characterized as "A trouble endlessly". And it is the truth - her negative consequences remain so far and for a long time, will long constitute the danger.

The Chernobyl accident, over 30 years later, remains the continuing accident. The tragedy without completion dates. The same can be told about the Chelyabinsk atomic enterprise and accident on the Japanese nuclear power plant in 2011.

Accident... This sad definition belongs to all cases of radioactive effects from various sources of artificial radiation. Radiation problems, in particular, her impact on the person, are one of the most urgent for the whole world. Scientific world-class and most authoritative institutions of the developed countries take great interest in the solution of their many questions. Influence of radiation on a human body is intensively studied more than 60 years. This baton will be passed to many subsequent generations.

Long-term observations are made, have to and will be always is carried out over the persons who have undergone different sources of radiation exposure (radiotherapy, accidents on the NPP, military exercises, etc.) and their descendants. As a result of such observations, despite some discrepancy of information, the established facts defining a straight line and the mediated radiation role in the emergence of a number of changes and diseases are established.

Treat them:

- cancer of thyroid gland which is especially often arising at children of younger age;
- solid cancer of various localization (lungs, chest gland, stomach, genitals);
- genetic diseases and mutations which frequency and weight are inversely proportional to pregnancy terms:
- bone marrow diseases (leukemia, anemia, etc.);
- violations of activity of immune system (allergy, immunodeficiency, autoimmune violations);
- increase in prevalence of diabetes;
- specific changes in the central nervous system of the organic nature;
- cataract.

The value of radiation in the development of many diseases, including cancer of various localization, has been recognized by World Health Organization only in 1995, i.e. nearly 10 years later after the accident on the CNPP. This recognition was preceded by convincing proofs of scientific many countries.

The diseases of heart mediated by radiation, lungs, digestive tract, endocrine, uric systems, eyes, skin and its appendages, etc. differ in a persistent current, badly respond to treatment and aren't always easily diagnosed. In places of radioactive pollution indicators of the health of the person, it is expressed have worsened. Life expectancy of the population and birth rate has decreased, and the incidence, prevalence, mortality, and primary.

The diseases of heart mediated by radiation, lungs, digestive tract, endocrine, uric systems, eyes, skin and its appendages, etc. differ in a persistent current, badly respond to treatment and aren't always easily diagnosed. In places of radioactive pollution indicators of the health of the person, it is expressed has worsened. Life expectancy of the population and birth rate has decreased, and the incidence, prevalence, mortality and primary disability have increased.

Expansion and deepening of ideas of mechanisms of radiation damage are of very great importance. The Chernobyl accident differs from all other sources of radiation of the person. For the first time in the world, millions of people are exposed to the long influence of small doses of radiation. It is revealed that the end result of long the influence of small doses in development of cancer, congenital and hereditary diseases, is similar to that at radiation by high doses. The specified data are submitted extremely important because the number of the persons receiving small doses many times over exceeds the number of the persons who have undergone single radiation by high doses.

In this regard the position of some scientists which sometimes is widely duplicated by mass media "about myths of Chernobyl" i.e. safety of the happened accident is surprising. The data of the UN published at the end of summer of 2005 surprise. The optimistic information specified without the bases, negative is apprehended by persons and institutions which it is long and profoundly study consequences of radioactive effects on the person. It is impossible to overestimate the value of the determined consistent patterns of the damaging influence of radiation at her various characteristics (a dose, duration of radiation, a way of a hit to an organism). In-depth pilot studies and clinical observations have created a basis for receiving new medicines and ways of treatment of radiation damages. Indications for their application are specified and also side effects are defined.

Along with the creation of new medicines, radio projector properties of some plants, flowers, berries, etc. are studied.

The great value is attached to minimization of radioactive effects, complex methods of rehabilitation and social protection of victims. Actions for primary and secondary prevention of radioactive effects are offered and are constantly improved. The paramount value of protection of various sources of radiation contamination (cyclotrons, the NPP, radioactive waste, etc.) and need of her improvement is unanimously perceived. Methods of long-term observation over victims are developed. They, to some extent, differ from each other in the certain countries.

The Chernobyl accident became the platform of the international commonwealth. It is obvious to all that the Chernobyl accident has concerned not only three republics, and today the in depended on the countries – Ukraine, Belarus, and Russia.

The increase in a radiation background observed right after the accident in the European countries promotes the increase in the prevalence of cancer and other diseases. By means of the world community, the social security of victims improves, works hard the centers of rehabilitation for children and adults. Techniques of diagnostics, prevention, and treatment of patients, victims of radiation are regularly corrected and improved.

It is necessary to hope that all atomic accidents will serve as a lesson for the whole world, will help to prevent them and to reduce consequences.

Positive assessment is deserved by a role of the public of many countries and the International organizations (the UN, WHO, etc.) in the solution of many problems, in particular, in connection with the Chernobyl accident.

The radioactive effects on the person, unfortunately, are the proceeding tragedy. In this regard it is necessary to expand the contingents examine, being exposed in various time to radioactive effects. It concerns also persons, immigrating to other countries.

The opinion is represented lawful that on the NPP it is possible to judge consequences of radiation accidents after 50 years.

It is necessary to agree with the opinion of the outstanding American scientist John Hoffman: "To proceed from the worst - to do everything that consequences were minimum."

For minimization of consequences of radioactive effects, the active position of victims and their environment has important value. For this purpose, it is necessary to have an idea of radiation sources, methods of diagnostics, treatment, and prevention of radiation damages. In other words, the evidence-based literacy of a wide range of people, their active position, will promote certainly the success in the solution of many questions of a problem "Radiation and health." It is necessary to add to it that in training of doctors and other medical staff there are practically no data on medical radiology.

DICTIONARY OF TERMS AND CONCEPTS

Absorbed (biological effective)
- dose energy absorbed by fabrics

Aberration
- violations connected with changes of chromosomes

Adequate protection
- corresponding

Adenoma
- high quality tumor

Adaptogens
- substances reducing a stress

Allergy
- enhance able sensitiveness of organism to external to influences

Alternative medicine
- auxiliary methods of treatment (acupuncture, physical therapy, etc.)

Anatomy
- science about a structure of bodies and systems

Antibiotics
- medicines of antimicrobic action

Antimutogens
- substances, prevent mutations

Aplastic anemia
- caused by injury of bon marrow

Artificial radiation
- received as a result of artificial splitting of atom

ARS
- Acute radiation sickness

Atomic bomb
- powerful explosive device containing radioactive elements

Atrophy
- reduction of the sizes of cells of body tissues

Atomic energy
- result of splitting of an atomic nucleus

Audiometer
- device for definition of a hearing disorder

- inflammation of bilious ways

Antioxidants
- reducing the number of free radicals

Appendix
- worm-shaped shoot of a thick gut

Aplastic anemia
- caused by injury of bon marrow

Artificial radiation
- received as a result of artificial splitting of atom

ARS
- Acute radiation sickness

Atomic bomb
- powerful explosive device containing radioactive elements

Atrophy
- reduction of the sizes of cells of body tissues

Atomic energy
- result of splitting of an atomic nucleus

Audiometer
- device for definition of a hearing disorder

Biodosimeter
- device for definition of a biological dose of radiation

Ber
- radiation size unit

Becquerel
- radiation size unit

Biodosimeter
- device for definition of a biological dose of radiation

Biopsy
- lifetime studying of cells of body tissues under a microscope

Bon marrow
- hemopoiesis organ

Brach therapy
- beam impact directly on a tumor

Carcinogenesis
- chaotic and fast cell fission of fabrics

Caries
- disease of solid tissue of tooth

Catalyst
- accelerator of biochemical processes

Cirrhosis
- growth of connecting tissue

Cholecystitis
- inflammation of a gall bladder

Chromosomal aberration
- changes of structure of chromosomes

Chernobyl "AIDS"
- oppression of immune system caused by radiation

Chromosome
- structure of the genetic code (DNA) which is in a cage kernel

Collective dose
- radiation level multiplied by number of irradiated

Complementary medicine
- auxiliary methods of treatment (acupuncture, physiotherapy, homoeopathy of and other)

Congenital diseases
- arise during pre-natal development

Correlation
- compliance of processes

Computer tomography
- method giving an idea of the sizes and a structure of bodies

Cyclotron
- installation for receiving radioactive materials

Cytogenetic effect,
- connected with damage of cells of tissue

Cystitis
- inflammation of urinary bladder

Cytogenetic effect
- connected with damage of cells tissue

CNPP
- Chernobyl nuclear power plant

Deactivation
- clarification from radioactive materials

Detector
- device, it is mediated the defining radiation level

DNA (dioxynucleinic acid)
- main substance of a kernel of a cage

Dose reconstructive (restored)
- definition (radiation doses

Dissymmetry (radiometry)
- definition of a dose of radiation

Dysplasia
- wrong development of cells

Ecology
- science about the external environment

Electrocardiography
- record of currents of a muscular cover of heart

Elektroencefalografiya
- record of currents of a brain

Elimination
- removal

Endometrial
- inside layer of a uterus

Enterosorbents
- medicines adsorbing the substances emitted from blood in intestines

Encephalopathy
- brain disease

Erosion
- superficial ulcer

EPR
- electronically magnetic resonance diagnostics method

Erythrocytes
- red blood little bodies - carriers of oxygen

Genetics
- science about heredity

Genetic effect
- impact of radiation on posterity

Genome
- heredity unit containing DNA

Giperplaziya
- increase in the sizes of cells of tissue

Glaucoma
- increase in intraocular pressure

Growth hormone
- stimulates reproduction of cells of body tissues

Gynecology
- science about diseases of female genitals

Haemostimulinums
- agents improving a hemopoiesis

Hepatitis
- inflammation of a liver

Hyperthyroidism
- increase in function of thyroid gland

Genetics
- science about heredity

Genetic code
- "program" of a cage

Genetic effect
- impact of radiation on posterity

Genome
- heredity unit containing DNA

Growth hormone
- stimulates reproduction of cells of body tissues

Gy
- radiation size unit

Gynecology
- science about diseases of female genitals

Hepatitis
- inflammation of a liver

Hypophysis
- site of a brain coordinating functions of internals

Homeostasis
- constancy of the internal environment of an organism

Hyperbaric oxygenation
- saturation of an organism oxygen under pressure

Feto-platsentary issufficiency
- violation of blood circulation in a placenta

Immunostimulytors
- means of increase in functions of an immune system

Immunosuppression
- suppression of functions of immune system

Impotence
- violation of sexual desire at men

Induction
- transfer

Incorporated isotopes
- being in cages of bodies

Interferon
- medicine rendering against viruses

Induction
- transfer

Incorporated isotopes
- being in cages of bodies

Interstitial nephrite
- inflammation kidney interstitium

Ionization
- change of protons in an atomic nucleus

Ionizing radiation (radioactivity)
- energy which is formed as a result of an exit of electrons of an atomic nucleus

Isotopes
- same element with various quantity of protons

Izoflavona
- substances reducing the level of free radicals

KGB
- Committee for State Security

Keratosis
- growth of a caver of a skin

Latent period
- time between radiation and the beginning of signs of damages of an organism

Leukemia
- blood cancer

Leucopenia
- decrease of quantity of leucocytes in a blood

Leukocytes
- white blood cells

Lipids
- kind of the fats participating in biochemical reactions

Liquidators ("Chernobyl veterans")
- persons who were taking part in recovery from the accident on the CNPP

Lymphoma
- cancer of lymph nodes

Magnitude
- size characterizing energy at an earthquake in the form of seismic waves

Mammography
- X-ray research of chest gland

Markers
- specific indexes of substances

Mediators
- transmitters of biological processes

Medical genetics
- science about hereditary disease

Mediators
- transmitters of biological processes

Medical genetics
- science about hereditary diseases.

Melanoma
- malignant skin tumor

Metastasis
- spread of a malignant tumor to various organs

Mikrotsefaliya
- reduction of the sizes of a brain

"Mongoloid" type
- changes of usual outlines of the person

Monitoring
- constant control

Mortality
- number of the dead in a year on 100 thousand population

Mycoplasmas
- special species of microorganisms

Myocardium
- muscular wall of heart

Myocarditis
- myocardium inflammation

Mutation
- genetic change of hereditary

Natural radiation
- environment radiation

Nephrology
- science about diseases of kidneys

Nephrite
- inflammation of kidneys

Nephropathy of pregnant women
- disease of kidneys at pregnancy

No threshold dose
- hypothesis excluding harmlessness of small doses of radiation

NPP
- nuclear power plant

Oncology
- science about malignant tumor

Orthopedics
- science about diseases of bones and joints

Osteoporosis
- decrease in density of bones

PAES
- floating nuclear power plants

PAES
- partnership for Environmental Sustainability

Phobias
- fears, fear

Penetrating radiation
- passing through integuments in inside organism

Perekisny oxidation
- stage in exchange of fats

Pericarditis
- inflammation of a pericardium

Phitotherapy
- treatment by herbs

Plasma
- liquid part of blood without uniform elements

Platelets
- blood plates participating blood coagulation

Placenta
- location of an embryo providing communication with the pregnant woman's organism

Pneumosclerosis
- growth of connecting tissue in lungs

Prodrm
- period preceding the beginning of a disease

Proktit
- inflammation of a mucous rectum

Proliferation
- reproduction of cells of body tissues

Prostatitis
- inflammation of a prostate gland

Psychotropic drugs
- exert impact on a brain

Radiation
- emission of power particles atoms

Radioactivity
- energy which is formed as a result of splitting of atoms

Radio diagnosis
- use of radiation for diagnostics

Radiology
- science about radioactive materials

Radio medicine
- section of medicine including radio diagnosis and radiotherapy

Radio therapist
- expert who is carrying out radiotherapy

Radiotherapy
- treatment by isotopes

Radioisotope
- element with an unstable atomic nucleus

Radio induced diseases
- connected with radiation exposure

Radio protectors
- substances reducing influence of radiation

Radioisotope scanning
- method of graphic registration and distribution intravenously entered isotopes

Radio patronage
- complex of the actions reducing influence of radiation

Radio resistance
- resistance to influence of radiation

Radio sensitivity
- increased perception of radiation

Radiophobiya
- fear of radiation

Rehabilitation
- complex of actions directed on recovery of health

Reconstructive dose of radiation
- determined through various terms after radiation by indirect methods

Reparative enzymes
- promote restoration of functions of cells of body tissues

Reproduction
- reproduction of posterity

Refraction
- violation of light refraction in an eye

Respiratory ways
- airways

Retrospectively
- after the incident

Sepsis
- bacterium blood poisoning

Sclerosis
- growth of a cicatricles tissue

Solid cancer
- malignant tumor of ferriferous cages

Somatic effect
- influence of radiation on internals

Spondylosis
- disease of intervertebral cartilages

Stem cells
- replace or stimulate reproduction cells of body tissues

Syndrome
- combination of symptoms

Terratogennic diseases
- congenital disease which aren't transferred to the subsequent generations

Thermal photography
- determination in the speaker of temperature of a part of the body special instruments

Thymus
- organ of immune system located on a neck

Technogenic radiation
- connected with professional activities

Thyroiditis
- inflammation of thyroid gland

Thrombosis of vessels
- obstruction of their gleam

Transfusion
- intra vascular introduction of medicinal solutions

Types of radioactive radiations
- alpha, beta, gamma differ in wavelength and penetration

UN
- United Nations

Uniform elements of blood
- erythrocytes, leukocytes, platelets

ABOUT THE AUTHOR

 Neli Melman - doctor of medical Sciences, senior researcher at Kiev research Institute of urology and Nephrology.
Co-author of 11 books (manuals, monographs, handbooks, etc.), 150 scientific articles.

In November 1989 by the refugee family arrived in the United States.
For 15 years worked in the laboratory of molecular recognition, National institutes of health.
Co-author of 40 articles on current issues in molecular Popular science magazine in the USA in Russian (co-author).

- "Health effects of the Chernobyl accident" (2000).
- "American family medical encyclopedia" (2001)

The author of the popular science publications in Russian and in English:

- Radiation and health (2010)
- Socialization of Russian Jews doctors in the United States" (2007),
- Just the facts. Anti-Semitism is on the path of education and science"]. (2011)
- Jews in space exploration" (2013, 2017)
- On the socialization of Russian-speaking doctors in the United
- Endowments syndrome About savants (2017)

The life and work of N. Melman was positively assessed in the Newspapers "Senior Beacon", November,1998; and "Cascade"; the books "To know and remember" (S. Zolotarev, new York, 2002). Kleiner, A. I., "On the wave of memory", 2014, "the Jews of Russia in medicine and biology (1750-2010)" the Biographical encyclopedia. (E. Lyuboshitz, 2013, Jerusalem).

Member of two UN conferences on problems of Chernobyl catastrophe (2006 and 2007) and two International conferences.

Systematically examines and publishes in Newspapers and magazines materials on topical issues of radiation medicine.